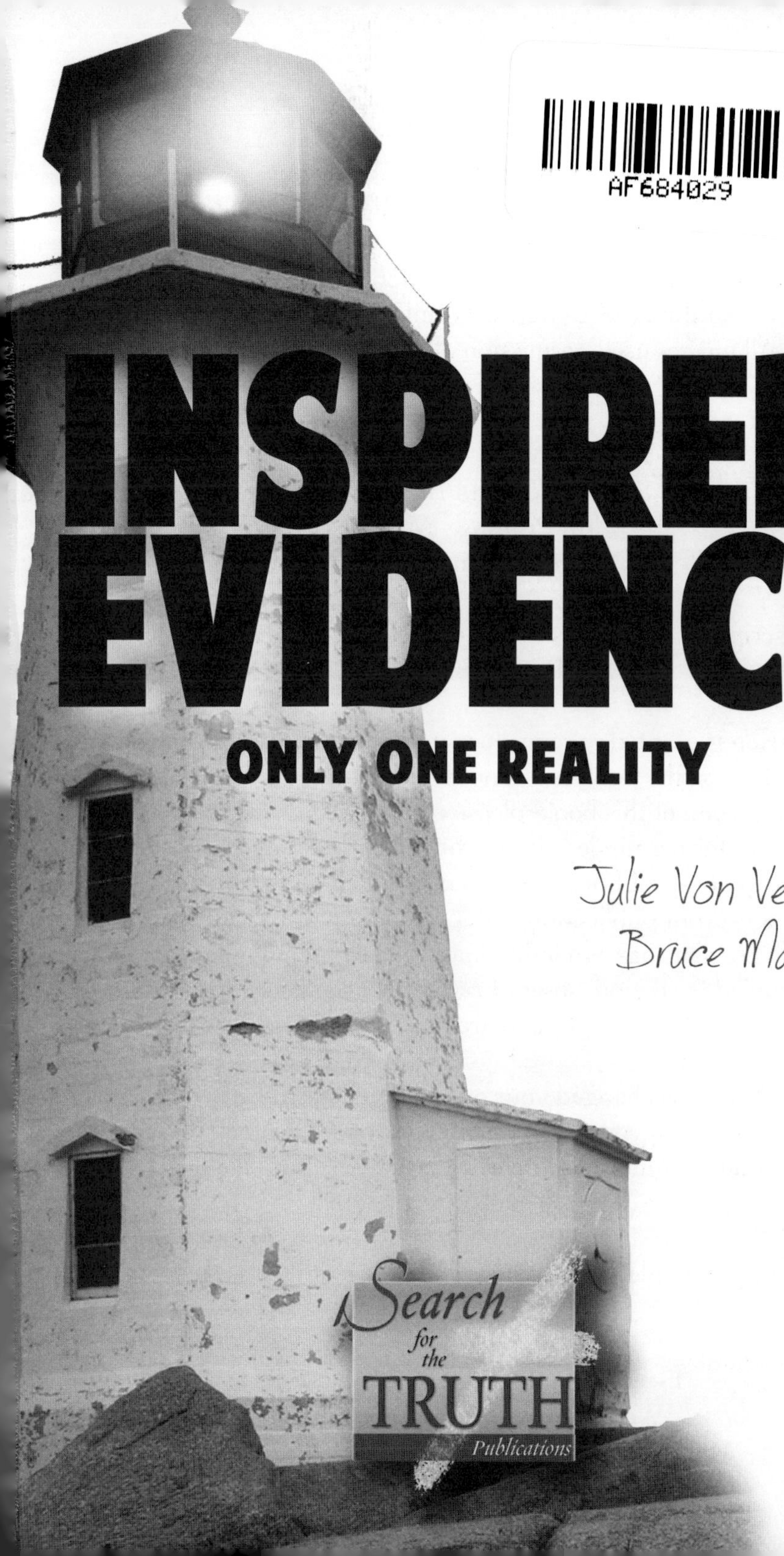

# INSPIRED EVIDENCE

## ONLY ONE REALITY

Julie Von Vett &
Bruce Malone

Search for the TRUTH Publications

Cover and interior design: Michael Malone

ISBN : 978-1-939456-24-3
Library of Congress Number: 2011912233
All images used permission of www.shutterstock.com
Printed in China

First Printing - December 2011
Revised Second Edition Printing - October 2012
Fifth Printing - May 2016
KJV only printing – Jan. 2017

All scripture quotations are taken from the King James Bible.
Printed in China



If this book has changed your outlook on life, strengthened your faith, or deepened your relationship with your Creator, please let us know. Low cost multiple copies are available using the order form at the back of the book.

May God bless you as you share the truth with others.

**E-mail** - truth@searchforthetruth.net
**Web** - www.searchforthetruth.net
**Mail** - 3275 Monroe Rd.; Midland, MI 48642

# DEDICATION

This book is dedicated to the pioneering work of Richard and Tina Kleiss to bring the evidence for creation to students. The Kleisses both served in our public school system – teaching two generations of students over a 32 year period. Their desire was for students to have an opportunity to view the evidence for a biblical worldview. Their devotional, *A Closer Look at the Evidence,* was the fruit of this vision. *Inspired Evidence* builds upon their foundation.

# ACKNOWLEDGEMENTS

The universe is an awe-inspiring place and attempting to use words to paint a picture of what God has done is a daunting task. In this volume the authors have humbly attempted to present a small part of the massive amount of evidence pointing to the trustworthiness of God's Word (epecially as related to creation.) In this effort we acknowledge the help of many others. Don, Jeri, and Aaron Slinger helped throughout the process with a complete editorial review, making many useful suggestions on creation-science related issues. The creative artistic design talents of Michael Malone were invaluable as he brought the text to life with visual images. Annette Schmidt proofread and corrected many errors in the first draft manuscript while Beth Kelch and family were instrumental in proofreading of the first edition. The second edition was greatly improved by the comments of Paul Even, Christine Stark, Charles Ramsey, Neil & Amy Legatt, Ed & Charlene Rosa, and Rick & Della Deighton who made countless improvements. Many ministry supporters also donated funds to make this book possible, we thank each of you. These faithful servants of the Lord, and many others, have our profound thanks. Mostly we acknowledge Jesus Christ, the creator, savior, and sustainer of all awesome things. Thanks to Jamie Walton for layout for KJV only devotional changes.

# FOREWORD

Two battleships assigned to a training squadron had been at sea on maneuvers in heavy weather for several days. The visibility was poor with heavy fog, so the captain remained on the bridge keeping an eye on all activities.

Shortly after dark, the lookout reported, "Light, bearing on the starboard bow."

"Is it steady or moving astern?" the captain called out.

The lookout replied, "Steady, Captain." This meant they were on a dangerous collision course with that ship.

The captain called to the signalman, "Signal that ship: 'We are on a collision course, advise you change course twenty degrees.'"

Back came the signal, "Advisable for you to change course twenty degrees."

The captain said, "Send: 'I'm a captain, change course twenty degrees.'"

"I'm a seaman second-class," came the reply. "You had better change course twenty degrees."

By that time the captain was furious. He spat out, "Send: 'I'm a battleship. Change course twenty degrees.'"

Back came the flashing light, "I'm a lighthouse."

God's Word is that lighthouse. No amount of human arrogance, popular opinion, or scientific theorizing will alter its eternal truths. The universe did not create itself; creation is recent; life does not come from non-life; plants and animals do not transform themselves into higher life-forms; there has been a globe-covering flood in Earth's recent history. The Bible is crystal clear on these facts and scientific observations, and when properly interpreted, confirms each one.

There can only be one truth, and those who oppose that truth will ultimately find themselves shipwrecked upon the unchanging rock of reality.

# INTRODUCTION

There are only two possibilities for explaining our existence.

**~ either (A) ~**

Some sort of simple organism turned into a more complex organism…which turned into some sort of marine creature…that turned into some sort of land creature…that eventually turned into people…

**~ or (B) ~**

The vastly different groupings of creatures (including people) were created fully formed and functional – yet programmed with the information needed to vary widely within a given "kind" of creature.

Those who believe the truth is a blend of the two possibilities often teach that God used some sort of evolutionary process or enormous periods of time (option A). By promoting huge time periods, they are essentially accepting that the Bible cannot be understood in a clear, straightforward way. This compromise position eventually deteriorates into the belief that there is no evidence from the study of nature for the existence of God.

Saying God used huge time periods to create life is in practice no different than accepting evolution as a fact. Thus this compromise position is really just a subset of position A.

*Yet there cannot be two realities.*
*One of these options is true, the other is fantasy.*
***Truth is that which conforms to reality.***

What can science tell us about truth? Basically, science cannot determine truth, but careful observations can determine which of two options is ***least*** likely to be true. This devotional showcases the observations which are systematically ignored, suppressed, and hidden from students as they are indoctrinated toward believing that option A is the ***only*** possibility.

As you read page after page of this devotional, each containing evidences from different areas of science, keep the biblical framework in mind. Ask yourself, "Does the evidence on this page best fit a biblical viewpoint or an evolutionary viewpoint (enormous time periods & slow modification)"?

So, what is the biblical viewpoint? A straightforward understanding of the Bible acknowledges five **REAL** events of history which should be supported by the observations of science:

1. The recent creation of the universe, the earth, and very diverse forms of life (including mankind). *[biological sciences]*
2. The rebellion of mankind resulting in the entrance of death, disease, and genetic deterioration of creation. *[genetics/human psychology]*
3. A global, world-restructuring flood upon the earth approximately 4600 years ago. *[geological sciences]*
4. A dispersion of mankind across the globe via different language groupings. *[linguistics/archeology/world history]*
5. The arrival of God upon the earth in human form to take the penalty of death upon Himself as a demonstration of ultimate love. *[philosophy/salvation/theology]*

Enjoy the awe inspiring evidence from these fields of knowledge. We believe you will be encouraged and stirred by how well scientific observation confirms God's Word. But most importantly, share this book with others. The primary weapon of those devoted to removing any relevant belief in God from our culture is the removal of creation from public discourse. They have used the courts and school system to completely shut down any discussion of the evidence for option B (creation). Yet, we still have the freedom to bring the truth to others (in essence going around the blockade to truth which our education system has become) by sharing this book with others.

***There can only be one reality***

***– share that reality with others.***

# ABOUT THE AUTHORS

Julie Von Vett is a well-sought–after creation speaker and has been teaching creation science at churches and community groups full time since 2003. Prior to that, she home educated her children, Annette and Caleb, for 10 years through high school. After attending worldview classes at Summit Ministries in 1992, she began teaching year-long worldview classes to high school students and parents. She realized that the foundation of a biblical worldview begins with creation, Genesis 1-11. She received her B.A. from St. Olaf College. Julie and her husband Vince live in Minnesota; in their free time they enjoy exploring the geology out West and digging up dinosaurs.

Bruce Malone has spent over 30 years bringing the scientific evidence for creation to churches and colleges at seminars throughout the United States and 14 foreign countries. He has authored 7 books on creation with 250,000 books in print and a creation video curriculum; served as adjunct speaker for the Institute for Creation Research; is an associate speaker for Logos Research Associates Inc.; and is a commissioned worldview speaker under Chuck Colson's Centurion program — bringing a recent Biblical creation understanding to worldview issues. Bruce has 27 years of research experience with the Dow Chemical Corporation but left in 2008 to serve as Executive Director of *Search for the Truth Ministries* with the vision of "Awakening Hearts and Minds to Biblical Truth." This organization widely distributes Bible-affirming creation materials to students and prisoners. Bruce has a B.S. degree in Chemical Engineering from the University of Cincinnati and holds 17 patents for new products with the Dow Chemical Corp. Bruce and his wife Robin have been married since 1983 and have 4 grown children - Michael, Marc, Margaret, and Matthew and 4 grandchildren. They reside in Midland, Michigan.

# JANUARY

*When there is hope for the future,*
*there is power in the present.*

**- John Maxwell** (1947 - present)
Christian author, speaker, and pastor

*Happy is he that hath the God of Jacob for his help, whose hope is in the LORD his God.*

**- Psalm 146:5**

The evolution/creation debate is similar to debating the existence of air. No matter how much evidence we put forward as to the existence of air, the person refusing to acknowledge air's existence would reject the evidence and propose some alternative explanation. Yet, the person denying the existence of air is breathing air and using air to transmit his denial of air's existence.[1]

In a similar way, evolution is simply a story to explain the existence of everything while denying the existence of a Creator. Those who believe in evolution make truth claims (random processes such as "mutations filtered by natural selection" created all biological life); yet, the very existence of truth depends upon the Biblical concept that truth exists. IF we are just gas molecules that turned into stars…that turned into elements…that turned into organic molecules…that turned into single-celled organisms…that turned into fish…that turned into land animals…that turned into ape-like creatures…that turned into people…then truth is a nonsensical concept.

How could hydrogen gas learn to think? How could truth be ***true*** if it happened by accident or chance? Why should one activity be considered wrong and another activity be considered right? Thus, those claiming that evolution is THE truth, in reality, have to rely upon a Biblical concept of the existence of truth to make their argument that the Bible cannot be true.

**Wisdom crieth without; she uttereth her voice in the streets:**

**21She crieth in the chief place of concourse, in the openings of the gates: in the city she uttereth her words, saying…**

**~ Proverbs 1:20**

The credibility of a truly biblical worldview hinges on one key event of history – the reality of a worldwide flood. If this event was simply a local flood, then the enormous rock layers of the Earth must have taken enormous periods of time to form. Therefore, the geologic record of death, disease, extinctions, and bloodshed is a record of God creating a world full of tragedy – before mankind ever appeared. If a global flood has not happened – God is evil.

However, if this flood really was global in extent and the consequence of a righteous God judging the actions of sinful mankind; it would have created the geology of our planet rapidly, and billions of years of Earth history simply disappear. Furthermore, God promised to NEVER send another flood like this upon the Earth – yet, there have been thousands of floods throughout Earth's history. So, either God is a liar who cannot be trusted, or the flood of Noah was very, very different than any subsequent flood.

## INSPIRED BY GEOLOGY

January 2nd

Noah's flood was global in extent! Much of the geology we see today was formed during or shortly after the Flood. By denying this flood, the geology of our planet and the timeframe of this geology is misinterpreted. This fact is central to a correct understanding of human history, the trustworthiness of God's Word, and the charactor of God.

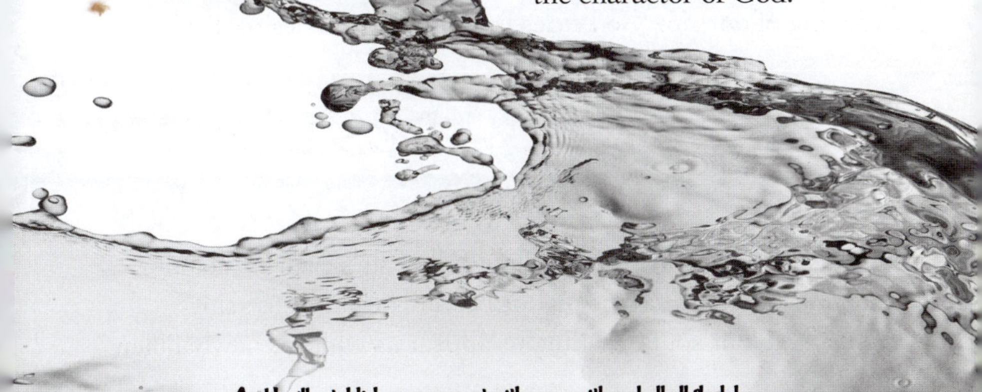

**And I will establish my covenant with you; neither shall all flesh be cut off any more by the waters of a flood; neither shall there any more be a flood to destroy the earth.**

**~ Genesis 9:11**

Have you ever gone outside on a sunny winter day and been nearly blinded by the brightness of the snow? Snow reflects nearly 90% of the sun rays landing on it. This keeps the snow cooler, resulting in a slower melting rate. Why would God want snow to stay cold longer? Keeping the snow cool is important because it stores much of the water we need for later use. When snow does melt, it is usually gradual, making flooding minimal. As the snow melts, it slowly percolates into the ground, making the ground ready for spring planting.

God designed snow to hold the water until it needed to be released. Water stored as snow protects the Earth from massive erosion that would take place if all moisture fell as rain. Snow acts as a fluffy insulating blanket for plants and creatures underneath the snow, shielding them from the harsh winter temperatures. It can be 20 degrees Fahrenheit warmer near the ground under snow than in the air above the snow.

Falling snow cleans the air. Every snowflake captures floating dust and other particles as it begins to crystallize and falls to the Earth. You may have noticed after a snowstorm that when you take a deep breath, the air smells clean and crisp. It is almost like the falling snow vacuumed the air for us!

The whiteness of snow is also a wonderful metaphor for God's cleansing of our sin. God cares for our soul and desires for it



to appear as white as snow and not full of sin's crimson color. Accepting Jesus' sacrifice on our behalf is what brings this cleansing to our soul – making us appear as white as newly fallen snow.

**Come now, and let us reason together, saith the Lord: though your sins be as scarlet, they shall be as white as snow; though they be red like crimson, they shall be as wool. - Isaiah 1:18**

Have you considered the beauty of snowflakes? A snowflake starts as a tiny crystal that forms around a speck of dust. As the water droplet freezes, the crystal starts to develop its six sides. As moisture is drawn into the crystal, the six branches begin to grow. As each snowflake is blown around in a cloud, it encounters different temperatures and moisture levels, resulting in no two snowflakes looking alike.

If it is extremely cold, for example below -40°F (-40°C), diamond dust snowflakes form. As the temperatures increase, different shapes will form: dendrites, prisms, or hollow columns. Researchers are still mystified why the shape of snow crystals varies widely with temperature. Humidity also is a factor; the more humidity present, the more complex the design.

The average complex snowflake is made up of 100 ice crystals. The number of ways these crystals can be arranged is staggering; the possible arrangements of these 100 crystals is greater than the total number of atoms in the universe. Thus, it's unlikely that any two complex snow crystals since the Flood of Noah's time have ever looked completely alike. Every snowflake is truly unique.

It is also interesting to compare man-made snow from snow making machines to "God-made" snow. Man-made snow does not have any design but looks like rough gravel. With God's snow, each snowflake has a unique design showing His creativity. God

# INSPIRED BY DESIGN

JANUARY 4TH

is still giving us new designs today. The next time it snows, catch a snowflake on your dark mitten and look to see the new design God has placed before your eyes.

**He giveth snow like wool: he scattereth the hoarfrost like ashes.**

**~ Psalm 147:16**

Have you considered the penguin? It is a bird, but it cannot fly. Why not?

The penguin dives for fish in the oceans. This bird does not have hollow, light bones but solid heavy bones with wings that act like paddles, so the penguin is equipped to swim. To swim and dive in the oceans with hollow bones would be like wearing a life jacket while diving to the bottom of a pool – very difficult, if not impossible. So, the great Creator gave the penguin solid, heavy bones which enable him to dive deep into the ocean to catch fish.

## January 5th — Biology Inspired By

There is no known link between penguin wings and bird wings. After all, how do you slowly change an intricately re-enforced, light, hollow bone structure into a heavy, solid bone? How do you slowly transform a wing meant for flying into a wing perfectly designed for swimming? No transitions are known, and they are even hard to envision.

Penguins also sport a fancy tuxedo. This tuxedo allows the swimming penguin to disappear in a flash when a predator spies him. Leopard seals, fur seals, sea lions, sharks and killer whales are the main threats for penguins in water. When a penguin swims, the black side of his tuxedo is up. If a predator is swimming above the penguin and looks down, the black penguin tuxedo blends in with the blackness of the ocean. If the predator is swimming below the penguin and looks up, the white part of the penguin tuxedo blends in with the light surface of the ocean, again making it difficult for the predator to spot the penguin. God has designed the penguin for successful living in the water. When you see design and purpose, you know it didn't happen by chance. There is a designer and that designer is God.

***O Lord, how manifold are thy works! in wisdom hast thou made them all: the earth is full of thy riches.***

***~ Psalm 104:24***

Did you know that penguins wear sunglasses? Penguins that live in the Antarctic experience intense glare from polar sunlight. Yet, they have clear vision. Why? Within their retina are colored droplets of oil, and these droplets filter out blue and ultra-violet colors which allow penguins to experience no glare and enhanced visual sharpness. Researchers are now copying the penguin's orange-colored filters to produce improved welding masks. Today's masks are orange in color rather than the

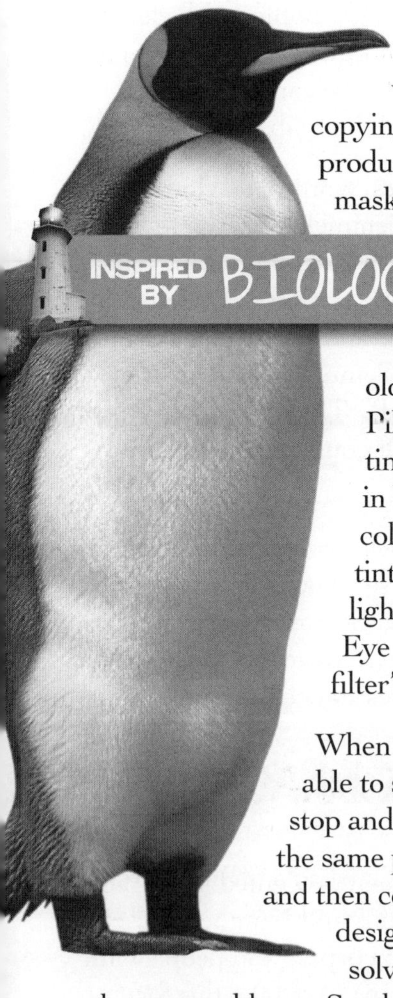



older dark shields which obscured vision. Pilots, skiers, and sailors all don orange-tinted sunglasses and find improved vision in bright sunlight, fog, or haze. Why is the color orange a useful light filter? Orange-tinted sunglasses reduce or eliminate blue light, which is a major component of glare. Eye damage is also minimized by the orange filter's blockage of the UV light.

When we face a problem, such as not being able to see well in bright sunlight, we should stop and think: "Does something in nature have the same problem?" Observe how God solved it, and then copy God's solution! God is the master designer, the master problem solver. He solved the polar glare problem, and He can solve our problems. So, the next time you see a penguin, say "Hey, cool sunglasses!"

***Great are the works of the Lord,***
***they are pondered by all who delight in them.***
***~ Psalm 111:2***

## January 7th BIBLICAL ACCURACY INSPIRED BY

Ignaz Philipp Semmelweiss, a medical doctor, worked in a hospital that had two wards, one run by midwives and the other run by doctors. In the midwives' ward, 4 out of 400 patients died, while in the doctors' ward, 100 out of 400 died. What was the difference? He investigated and found that doctors did not bother with cleanliness. Doctors would routinely leave the morgue without washing and then go directly to the patient ward. In the early 1800's, doctors took pride in how blood-encrusted their lab coats became.

The midwives, in contrast, did no autopsies, wore clean clothes, and frequently washed their hands. Dr. Semmelweiss saw the difference and instituted hand washing with soap and water. The death rate in the doctors' ward immediately dropped. The other doctors thought this was just a coincidence and rebelled against the dreadful burden of washing their hands. Semmelweiss was fired for causing discord among the medical staff. The hospital threw out the basins with the soap and water. Of course, the death rate climbed back to record highs.

Much later, in the 1860's, doctors finally acknowledged that hand-washing and cleanliness did prevent disease. Thousands of people needlessly died between the time when Dr. Semmelweiss solved this problem and his solution was finally acknowledged. These deaths were caused by pride, ignorance, resistance to change, and stubbornness. These deaths could have been prevented if the Bible had been believed. In the books of Leviticus and Numbers, strict rules are set forth concerning contact with corpses and people with disease (Numbers 19:13-17, Leviticus 15:13). We take this for granted today, but God revealed these truths over 3,000 years ago. Whenever the Bible touches upon scientific subjects it is accurate. Instead of washing our hands of God's Word, we need to apply all that it says to our lives – even when it comes to washing our hands!

**And whomsoever he toucheth that hath the issue, and hath not rinsed his hands in water, he shall wash his clothes, and bathe himself in water, and be unclean until the even. ~ Leviticus 15:11**

Wisconsin is home to a very important fossil find – the jellyfish. So what's the big deal about a fossil jellyfish?

In Central Wisconsin, near the town of Mosinee, there is a rock quarry. But this is no ordinary quarry; it contains many thousands of fossilized jellyfish. Why are they there? In the oceans, jellyfish are squishy blobs of clear gel. How could squishy jellyfish fossilize? Generally, hard substances become fossils – not soft substances. Also, if you have ever been to the beach, you have probably noticed seagulls eating things washed up on shore, maybe even a jellyfish. When animals wash up on shore, there are predators around to eat them – they simply do not lie on the shore waiting to be fossilized. Furthermore, when a jellyfish washes up on shore, it pumps its bell, trying to get back to the water. This will leave behind little rings in the sand. In this rock quarry, no evidence of fossilized rings around these jellyfish exist. A jellyfish is 96% water and would dry out and shrink if exposed to air. There is no fossil evidence that these jellyfish have shrunk.



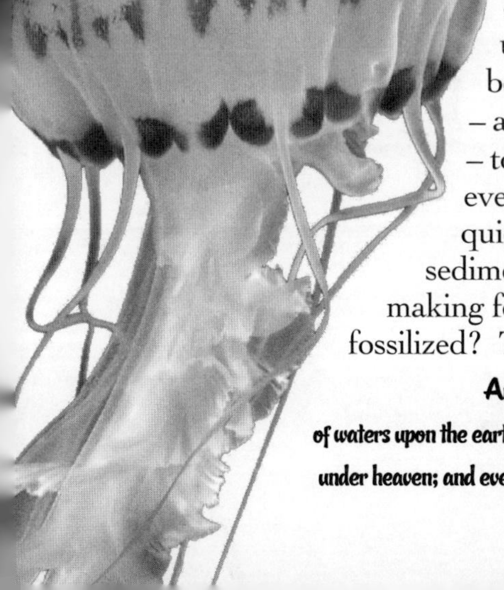

Moreover, these jellyfish were not found in just one layer but were found buried in seven layers of the quarry over a thickness of 12 feet.

So, what does all this evidence tell us? These jellyfish had to have been covered and fossilized quickly – and not just once – but repeatedly – to form the many layers. What event in history would have fast, quick coverage with mineral-filled sediment and no oxygen (the recipe for making fossils), so the creatures could be fossilized? The Flood of Noah!

**And, behold, I, even I, do bring a flood of waters upon the earth, to destroy all flesh, wherein is the breath of life, from under heaven; and every thing that is in the earth shall die. ~ Genesis 6:17**

You may think that lightning only brings fear and sometimes destruction. What good is lightning? Plants need nitrogen to grow green and healthy, and our atmosphere is 78% nitrogen. Yet, plants cannot take the nitrogen they need for growth directly from the air. Instead, our Master Designer has provided an alternative means of getting that nitrogen into the plants–through lightning.

Lightning actually converts the nitrogen in the air to nitrous oxide, which falls to the ground and enters the soil where it can be used by plants. It is estimated that electrical storms produce one hundred million tons of usable nitrogen for the Earth's plants each year. You could call lightning a super fertilizer maker! So the next time



there is an electrical storm, know that God is just turning the lawns green.

*... He maketh lightnings for the rain...*

*~ Psalm 135:7b*

People often complain about the weather here on Earth, but after reading the weather report for other planets, we might want to change those complaints to praises.

- **Mercury:** Extreme temperature swings today! A scorching 870°F (466°C) in the sun, -300°F (-184°C) in the dark, and an atmosphere of unbreathable hydrogen and helium.
- **Venus:** Skies overcast with heavy clouds and periodic thunderstorms of sulfuric acid. Temperatures day or night of 900°F (842°C). Air pressure of 100 times greater than on Earth. You will be suffocated, crushed, roasted, and corroded.
- **Mars:** Dust storms are approaching. Stay indoors since there



is not enough oxygen to breathe. Lows of -100°F (-73°C).
- **Jupiter:** Today there is an extreme smog alert. The air will smell of rotten egg gas because the air is full of poisonous hydrogen sulfide. Be on alert for cosmic radiation, falling meteors, and magnetic storms. The "red spot" hurricane continues to roar.
- **Saturn:** -216°F (-138°C). Winds blow at 1,118 mph. Gas planet, so there is no ground to stand on. Clouds made up of ammonia crystals.
- **Uranus**: -355 °F (-215°C). Poisonous atmosphere.
- **Neptune:** Winds will blow non-stop at 1,300 mph. The days will be dark since Neptune receives only one-tenth percent of the amount of the sunlight as compared to Earth.
- **Moon:** No rain. No wind. No clouds. No blue sky. Afternoon temperature 253°F (123°C). Overnight temp. -384°F (-231°C).

Compared to other places in our solar system, Earth's weather is just right. There really is no place like home!

**God himself that formed the earth and made it; he hath established it, he created it not in vain, he formed it to be inhabited.. ~ Isaiah 45:18**

Leaves are little green machines running on air, water, sunlight, and a few minerals. These miniature solar panels are involved in an incredible process called photosynthesis. When sunlight falls on the green cells in the leaf, chemical changes take place, like the splitting of the molecules of water into oxygen and hydrogen. Water is extremely stable, so causing water molecules to split apart into oxygen and hydrogen is a very complex process requiring lots of energy input. The oxygen is released into the air we breathe, while the hydrogen is used to make sugars in the leaf as it combines with carbon dioxide from the air. This process involves more than 70 complex chemical reactions. The sun provides the energy needed to drive this process, but how did a plant cell ever learn to harness this energy? Pure energy destroys biological processes; it does not create order.

Photosynthesis is so complicated that scientists have been unable to duplicate the entire sequence of events in the lab, yet it is done automatically in a leaf. If evolution was true, how did this come about? What would be the step-by-step process which allowed the complicated chain of reactions to have fallen into place? What was the step just before a full process of photosynthesis developed in a plant's life? How could that pre-plant have survived to become the first plant capable of photosynthesis?

Until these types of questions are satisfactorily answered, evolution is just story-telling and faith (evolutionism), not science. Plants needed photosynthesis before they could exist and our Master Designer engineered this incredible 70-step process.

**He that trusteth in his riches shall fall:**
**but the righteous shall flourish as a branch.**
**~ Proverbs 11:28**

Blood is the most amazing product transportation system ever devised. Arteries and veins are the small flexible tubes that carry blood to and from our cells, delivering oxygen and removing waste. Our blood is mostly made up of plasma that carries the other solid materials: red blood cells, white blood cells and platelets. Red blood cells carry oxygen, white blood cells fight infections and platelets plug leaks in the veins and arteries.

## INSPIRED BY ANATOMY

### JANUARY 12TH

The average person knows very little about how his body works. Most people know more about the fluids in their car than the fluids in their own body. The 75 trillion cells in your body have many common characteristics. Each cell must breathe; each cell needs oxygen and food as fuel; each cell gets hot and needs cooling; each cell performs work and gives off exhaust gas and waste products.

Car engines have separate systems for each function; however, God has combined all of the required functions into one system in our body – the circulatory system. Blood carries food, oxygen and other required chemicals and minerals. It constantly moves required nutrients from processing sites to delivery points and transports waste from the cells to disposal plants. It even acts to surround and destroy invading enemies (like germs) and replace worn out or damaged parts of a cell. The motor of the circulatory system is the heart, a two cylinder pump, which keeps working even as we sleep, sit, or run a race.

Would we say that a car with all its systems happened by accident or chance? Car makers would not; they proudly display their name on the cars that they make. Think about how much more intricate is God's design of our circulatory system! Do you proudly give God credit for life's design?

**The life of the flesh is in the blood.**

**~ Leviticus 17:11**

In 1927 a major geological event happened off the coast of Newfoundland, Canada. Without warning, an earthquake triggered an underwater avalanche causing loose sediment on the continental shelf to slide deeper into the ocean. This mud raced down the slope along the ocean floor similar to wet cement flowing off a cement truck at 60 miles per hour. When it finally came to a halt, the new sediment layer had formed, covering 40,000 square miles to a depth of two or more feet.

During the 1960s, with the help of submarines, geologists were able to study this catastrophic deposit. Through this study, they realized that this layer formed a new turbidite rock layer (as they are called) and that these turbidites are a result of catastrophic water events – not slow and gradual processes. These common, widespread, uniform rock deposits were previously thought to have



taken hundreds of thousands of years to form, but because of what geologists discovered through the Newfoundland event, they now know these types of rock layers formed extremely rapidly. Many rock layers throughout the Appalachian and Rocky Mountains were actually formed by turbidities. More and more geologists are realizing that catastrophes formed the geological features of our planet.

The Flood of Noah's day would not have been a limited local flood but a violent catastrophe accompanied by earthquakes and volcanoes that completely restructured Earth's surface. Acknowledging this fact is controversial because it wipes out the huge time periods needed to explain life without God.

**And spared not the old world, but saved Noah the eighth person, a preacher of righteousness, bringing in the flood upon the world of the ungodly;**

**~ 2Peter 2:5**

One of the classic tests to determine if science is being properly applied is called the "white swan test". Suppose I wanted to believe that "ALL swans are white." In science this would be called my starting hypothesis. Could I prove this to be true by showing that all swans on a local pond are white? How about if all the swans in a given city were found to be white? Is it proven if I send e-mails to my friends all over the world, and they report back that all the swans they see around them are white? Have any of these observations proven my theory as scientifically true? Actually not. The only way to prove if any theory is true is to disprove it – in this case by finding a black swan. This is what is meant by science needing to be "falsifiable".

Molecule-to-man evolution is a belief system rather than a scientific theory because no evidence is ever accepted as a falsification of the belief. Evidence supporting evolution is exactly analogous to only looking for "white swans", while any "black swans" (which show evolution to be completely impossible) are simply ignored. These "black swans" include the impossibility of matter appearing from nothing, systematic gaps between very different kinds of creatures in the fossil record, the rapid build-up of mistakes upon the genetic code of living creatures, the majority of dating methods that indicate a recent creation, the multiple evidences for the rapid and recent deposition of the geological layers of the planet, the inability for complex multiple parts of an organism to appear simultaneously, and the impossibility of the information content of DNA to have made itself. This book is filled with these and many more "black swans".

These well documented observations, that support the reality that life was designed is exactly what we would expect if the Bible is true and the universe was recently and supernaturally created. Yet all of these contradictions of evolution are simply ignored as though the "black swans" do not exist.

***Trust in the Lord with all thine heart; and lean not unto thine own understanding. In all thy ways acknowledge him, and he shall direct thy paths.***

***~ Proverbs 3:5-6***

A species of millipede called apheloria has an almost unbelievable defense mechanism. Many millipedes roll up or give off a foul smell when attacked, but not this one. Each segment of this millipede's body contains special glands that make a chemical for its protection. When the enemy attacks, the millipede mixes this chemical with a catalyst. The resulting reaction produces hydrogen cyanide gas – the same gas used in Hitler's gas chambers to kill people!

Both the millipede and the enemy are engulfed in a cloud of this deadly cyanide gas. The enemy dies – while the millipede simply walks away! How did the millipede develop this gas? How did it develop immunity to cyanide gas at the same time? How many millipedes were eaten before they developed this type of defense? How many millipedes died by gassing themselves? Evolution would say all this happened by mutational accidents. The faith of evolution believers to explain everything via evolutionary changes is more unbelievable than the millipede's unbelievably complex defense mechanism! This amazing defense system is clearly the result of intelligent designing and planning.

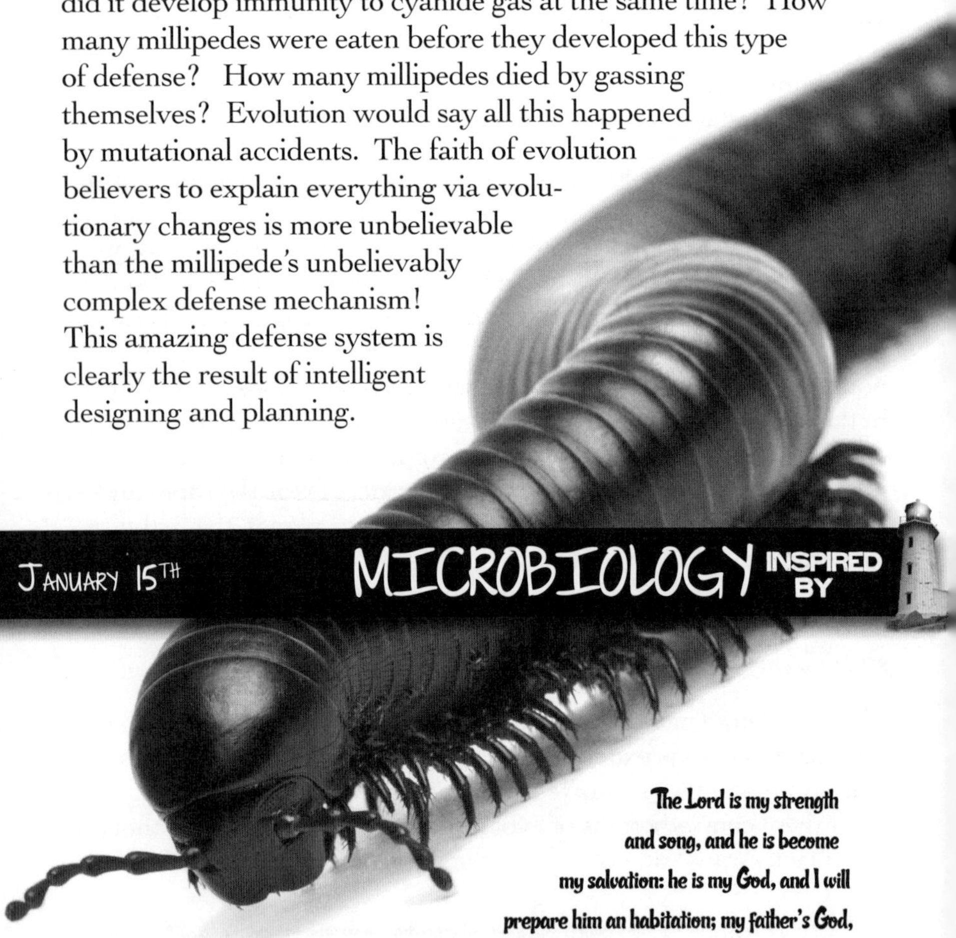



# MICROBIOLOGY INSPIRED BY

**The Lord is my strength and song, and he is become my salvation: he is my God, and I will prepare him an habitation; my father's God, and I will exalt him.**
**~ Exodus 15:2**

Did you realize that dinosaurs are mentioned in the Bible? In Job 40, we find a description of a behemoth. It feeds on grass like an ox (v.15), its tail sways like a cedar tree (v.17), its bones are like tubes of bronze and rods of iron (v.18), the behemoth ranks first



among the works of God (v.19), it hides among the reeds in the marsh (v.21). This behemoth is sometimes identified as a hippo or elephant by liberal theologians. However, neither of these animals fully fits the description. Instead, the behemoth sounds like an Apatosaurus (formerly called the brontosaurus). God was describing a creature with which Job was familiar. Since Job lived after the Genesis flood, this dinosaur did not disappear in the flood event. You will recall that Noah took representatives of all basic animal groups aboard the Ark, and that would have included the dinosaurs. They left the Ark and repopulated the Earth. God was just describing a huge creature that Job had seen.

Job 41 mentions a creature called the leviathan. Again, details of the creature are mentioned: its strong scales are his pride (v.15), its sneezes flashing forth light, out of its mouth go burning torches, sparks of fire leap forth, out of his nostrils smoke goes forth as from a boiling pot and burning rushes, his breath kindles coals and a flame goes forth from his mouth (v.18-21), when he raises himself up, the mighty fear (v.25). Liberal theologians, influenced by evolutionary thinking, attempt to identify this creature as a crocodile. It simply does not fit.[1] God is describing this creature, the leviathan, to Job, and it sounds very much like a seagoing plesiosaur with the ability to produce smoke or flames. This is not impossible since even the lowly bombardier beetle has been given this same ability.[2]

The people of Job's day were clearly familiar with both the behemoth and leviathan. Throughout human history and the Bible, these type of creatures were referred to as "dragons". Apparently, like many other creatures, they have become extinct since the time of Job and Noah.

**Canst thou draw out leviathan with an hook? or his tongue with a cord which thou lettest down?**

**~ Job 41:1**

Even the feces from baby song-birds reveal the design of God.

Baby song birds are always hungry, and the parent is continually feeding them, but what goes in one end…must come out the other! The nest could easily become a real mess. Imagine the bird nests filling up with bird excrement as the baby birds are pushed closer and closer to the rim by the rising sewage.

The presence of this fecal matter would not only be unhealthy but enable predators to easily detect their location. How did God



solve this problem? Disposable diapers. Each chick's fecal matter has a mucous membrane that surrounds it. The chick generally defecates within seconds of being fed. The parent then removes this fecal sac and deposits it away from the nest as it flies off to find additional food for the chicks. Shortly before the chicks fledge (fly away from the nest), they stop producing fecal sacs.

When you see tiny baby birds in their nests, think of how God has provided for their cleanliness and safety using the world's FIRST biodegradable, disposable diapers!

**I know all the fowls of the mountains: and the wild beasts of the field are mine.**
**~Psalm 50:11**

Have you considered that you might be wearing a souvenir from the Flood of Noah's Day? It's that amber necklace or earring. Amber is fossilized tree resin. Many pieces contain well-preserved land and aquatic critters such as insects, flowers, bird feathers, mammals, water beetles, and barnacles. Although we are not sure why amber seems to be concentrated in only certain parts of the world (like the Baltic regions), recent studies found that amber needed lots of water to entomb these critters. Water delays the process of "amberisation" (i.e. tree resin hardening). The tree resin stays stickier for longer periods of time underwater as compared to exposure in the air. With the amber being stickier longer, more critters get caught within the floating globules of tree resin. But how could both freshwater and marine critters become trapped in the same sticky tree sap? A catastrophic flood of worldwide proportions would have caused the uprooting of trees in enormous numbers, with waves causing copious quantities of tree resin to be released as billions of trees smashed against each other in the swirling currents. This sticky resin would have enveloped both aquatic and land organisms caught in the flood waters.

The next time you see someone wearing a piece of amber jewelry, try this conversation starter, "I see you are wearing amber. Have you checked to see if it contains the critters entombed during Noah's Flood?" Hopefully, you will never look at a piece of amber as just a piece of amber.

**The Lord sitteth upon the flood; yea, the Lord sitteth King for ever.**
**~ Psalm 29:10**

Imagine leaving your lunchmeat sandwich along a trail in the Grand Canyon during a family vacation and finding the same undecayed sandwich 50 years later. A ridiculous fantasy? Now imagine finding a chicken drumstick left on a rock in England by King Arthur 1000 years ago, yet the meat was still strangely soft and edible. Absurd? Imagine exploring an ancient tomb of Egypt that no one has entered for 4000 years and finding a plate of food as succulent and fresh as the day it was placed there. Magic? Yet all of these fantasies pale in comparison to what evolutionists need to believe about undecayed tissue finds within fossils.

In 2005, Dr. Mary Schweitzer found undecayed tissue, blood vessels, and cells within a fossilized T-rex bone[1] that she admitted should not have been there because it "violates everything we know about the laws of chemistry and physics". Amidst a storm of criticism, Dr. Schweitzer has proven that the tissue is indeed dinosaur tissue and found additional soft tissue in hadrosaur bones (assumed to have been buried 80 million years ago).[2] Meanwhile,

## January 19th — Paleontology Inspired By

Maria McNamara has identified muscle tissue (laying outside of the protective bone structure) from a salamander supposedly buried for 18 million years, and Johan Lindgren has found soft organic remains in the eye socket of a monasaur fossil supposedly buried for 80 million years. All of these time scales are ridiculous in the face of the soft tissue finds but make sense if the rock layers are the result of a recent worldwide flood.

God is making it absolutely apparent that a recent creation and a worldwide flood are a reality of history. His Word can be trusted as clear and straightforward. Dinosaurs were caught in a worldwide flood only about 4,500 years ago, and that is why small amounts of undecayed/unfossilized tissue remains.

**If I have told you earthly things, and ye believe not, how shall ye believe, if I tell you of heavenly things?**

**~ John 3:12**

Have you considered if there is intelligent life in outer space? This is a popular idea but only makes sense if viewed from an evolutionary viewpoint. If life evolved on Earth by natural processes, then intelligent life must exist elsewhere in space. But what does a



Christian worldview, based on the Bible, tell us about intelligent life in outer space?

During creation week, God created Earth on the second day. On Day 4, God made the sun, moon and stars (Genesis 1:16). Why did God make them? "For signs and for seasons, and for days and years (Genesis 1:14)". The Earth was center stage, and the sun, moon, and stars were created for helping us on Earth measure the passage of time (among other things). In the Old Testament, many passages distinguish between the heaven and Earth – always giving Earth a special place. Isaiah 66:1 says, "Thus says the Lord, the heaven is my throne, and the Earth is my footstool." The Earth is apparently a very unique spot in the universe.

But there are even more compelling theological reasons that point only to the earth being inhabited by intelligent life. In Romans 8:22, it states that the "whole creation groans" because of Adam's sin. When Adam sinned, the entire universe was affected, even outer space. Every year stars explode in supernova events throughout the universe. Now here is the problem. If there were intelligent life on other planets, then why would a just God afflict them with death because of what Adam did here on Earth? Death exists because if it didn't we would live forever separated from our Maker. Jesus overcame death and made a way for us to come back into fellowship with our Maker. The penalty for sin is death. Jesus paid that penalty with his death on the cross. If intelligent life existed elsewhere, those beings would also be affected by sin's action and subjected to death (just like stars are decaying). Yet they would have no opportunity for being saved. God wants all to come to salvation. Intelligent life exists only upon Earth.

**The heaven, even the heavens are the Lord's:**
**but the Earth has he given to the children of men.**
**~ Psalm 115:16**

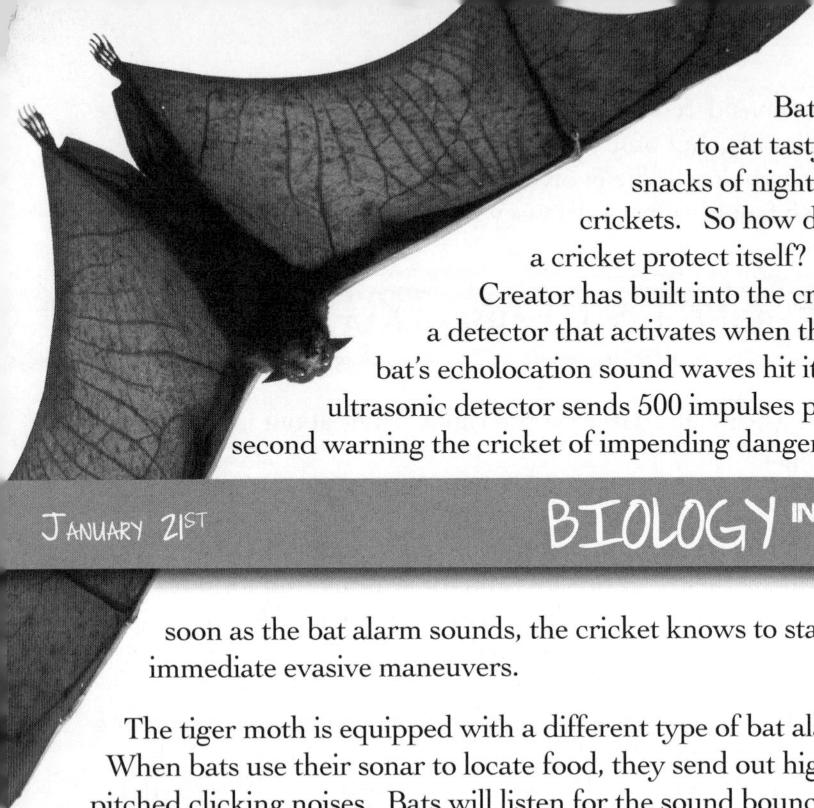

Bats love to eat tasty snacks of night-flying crickets. So how does a cricket protect itself? The Creator has built into the cricket a detector that activates when the bat's echolocation sound waves hit it. This ultrasonic detector sends 500 impulses per second warning the cricket of impending danger. As soon as the bat alarm sounds, the cricket knows to start immediate evasive maneuvers.

The tiger moth is equipped with a different type of bat alarm. When bats use their sonar to locate food, they send out high-pitched clicking noises. Bats will listen for the sound bouncing back. If the bat wants to investigate further, he clicks faster to hear more echoes. The bat clicks progressively faster as it "zeros in" on its target. Tiger moths can hear the bat tracking them. These moths will then try to confuse the bat by doing loop-de-loops and barrel rolls. If the bat continues tracking the moth, it will fold its wings and fall to the ground like a rock. Some of these tiger moths will even try jamming the bat's radar by sending out their own clicks.

Think about the incredible complexity of these defense mechanisms. How could a "simple" cricket or moth learn such behavior? How could it have developed the sound detecting equipment, mental processing capabilities and maneuvers found in a complex fighter jet?

**Bless the LORD, all you His hosts, you who serve him, doing his will. Bless the LORD, all you works of his, In all places of his dominion.**

**~Psalm 103:21~22**

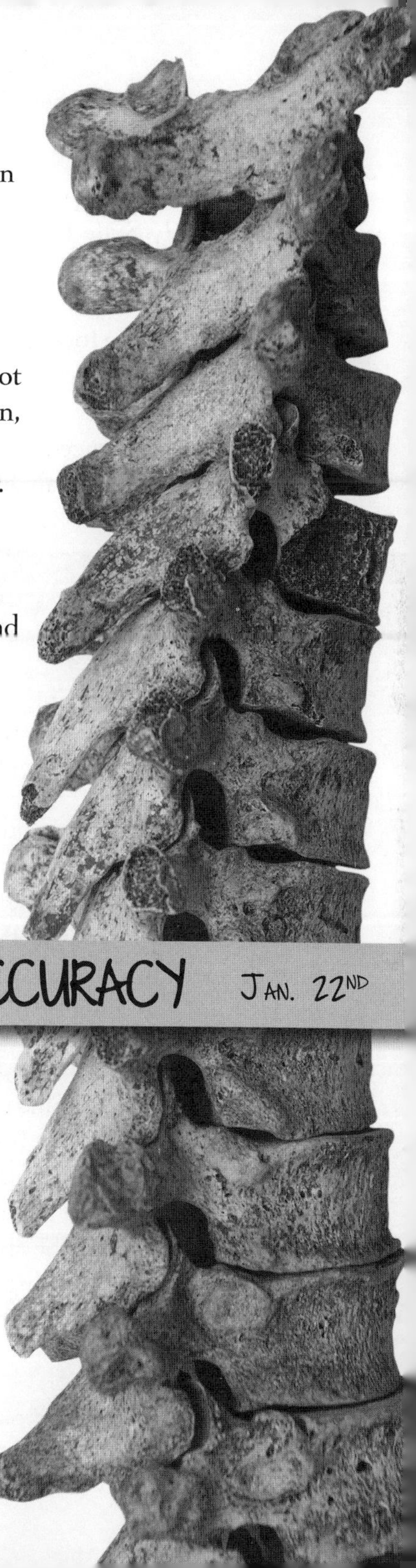

Many people have experienced upset stomachs which accompany worry, but did you know that unhappiness can even affect your bones?

A recent medical show was looking at the risk factors for osteoporosis. Some of the risk factors are being overly thin with a small skeleton, being a smoker, not drinking enough water, the lack of boron, or drinking alcohol. But scientists have also found depression to be a risk factor. It seems "depression causes inflammatory chemicals that suck away some of the bone". In other words, a depressed person does not have a merry heart...and this will affect their bones.

Essentially the same observation was made by King Solomon almost 3,000 years ago (see Proverbs 17:22). We are just now catching up! Of course, God has made us and wants to keep us healthy. Wherever the Bible speaks on



science, it is correct. The Bible is truly the inspired Word of God. So laugh a little and be joyful. It is good medicine for the bones.

**A merry heart doeth good like a medicine: but a broken spirit drieth the bones.**
**~ Proverbs 17:22**

How does one choose between the most popular ideas concerning the formation of the universe?

- The Big Bang Expansion (Boom and stretch)
- The Steady State Universe (Nothing new…same old, same old)
- The Forever Expanding Universe (And you thought diamonds were forever)
- The Cyclic Universe (The big crunch)

Some current evidence indicates that the universe is gently oscillating, enough to maintain equilibrium while preventing collapse.



This equilibrium is maintained by a gentle expansion and contraction of space. Almost like breathing! Even if this is true, where did everything come from, and how was this perfect balance achieved?

The Big Bang explains neither the origin of matter nor the perfect balance point achieved. A steady state universe could not exist forever, or all energy would be used up. The same problem eliminates the possibility of a forever expanding universe or a forever cyclic universe. All three choices violate the basic laws of physics and could not be true.

The correct answer is found in the Bible – God rolled out the heavens like a scroll. This reality accommodates all known observations, and the time frame fits the biblical model because during the rolling out process, space itself was stretched out. If this is indeed how the heavens were formed, the speed at which things were happening, including the speed of light, would not be limited to current values. Everything from background radiation to red shifts from distant starlight can be shown to fit nicely into this interpretation of cosmology.

**... and the heavens shall be rolled together as a scroll...**

**~ Isaiah 34:4**

Diamonds require a very high pressure to form, so they most likely formed some 200 miles below the Earth's surface. Geologists believe they were transported to the surface supersonically in violent volcanic eruptions. Diamonds are assumed to have formed 1-3 billion years ago because evolutionists need these extended time frames to explain life without God. In actuality, there is no reason to doubt that diamonds were compressed rapidly and recently during the Earth's formation.

Diamonds are made of carbon, so we can actually test their age with radiocarbon dating (commonly known as carbon-14 dating). The carbon-14 decay rate (or half-life) for this type of carbon is 5,730 years. After 5,730 years, half of any original amount disappears. After two half-lives, a quarter is left; after three half-lives only an eighth is left; after 10 half-lives, less than one-



thousandth is left. **Not a single carbon-14 atom should exist in any carbon source older than 250,000 years.**

Surprisingly, nationally certified radiocarbon labs have reported that diamonds have remaining carbon-14 amounts at over 10 times the detection limit. This is extremely convincing evidence that diamonds were indeed formed recently and rapidly. Diamonds are a biblical creationist's best friend!

Soaring some 18 feet off the ground, the giraffe is the tallest living land animal. Being so tall, giraffes have one of the highest blood pressures of all animals; they need two-foot long hearts to pump blood up those ten-foot long necks. Yet the brain is very sensitive to high blood pressure. Consider the giraffe taking a drink. He spreads his front legs and bends his head down. Can you imagine the blood rushing down his long neck, gaining speed and finally slamming into his brain? What a headache! Ah, but the Creator has built special design features into the giraffe. Within the blood vessels are check valves that slow down the rushing blood, and



when the blood finally does arrive at the brain, there is a "wonder net." This wonder net acts like a sponge, dispersing the blood around the brain.

Evolutionists believe these features happened by accident and chance. If this were true, imagine the giraffe getting a drink when the creature did not have the wonder net and check valves. The giraffe hears a lion, raises his head too quickly, and the blood rushes away from his head – causing him to black out. As a result, he is now lion food. But if he raises his head too slowly, the lion attacks, and again the giraffe is lion food. How many giraffes would have to be eaten until these two special design features evolved? How could they have just happened at the same speed at which his neck was miraculously lengthening? Evolution requires an enormous amount of faith and story-telling.

The check valves and wonder net were needed right from the beginning in order for giraffes to survive. God created the giraffe on Day 6 of creation week with all of these features. These design features cry out, "There is an Almighty God who made this!"

**O LORD, how great are thy works! And thy thoughts are very deep.**

**~ Psalm 92:5**

Since a giraffe has one of the highest blood pressures in the animal kingdom (needed to pump blood way up to its brain), why doesn't the blood pool down in his skinny legs? The reason is that the giraffe has been designed with tight-fitting skin. When a giraffe walks, the muscle movement within that tight skin helps pump blood up and out of the legs. You could say a giraffe is wearing support hose, the kind that you wear in the hospital to prevent blood from pooling in your legs.

## INSPIRED BY BIOLOGY

JANUARY 26TH

Giraffes also walk differently from most animals. Most animals walk by putting the left front foot forward at the same time as the right back foot, sort of diagonal. Giraffes walk by putting the left back foot forward along with left front foot. The giraffe is long-legged and short-bodied; if they walked like most animals, they would be hitting their legs against each other and stumbling. God, in His wisdom, programmed the giraffe to walk the way it does from the beginning.

Also notice the giraffe's long tail. Why doesn't such a long tail get caught on brushes and tree branches when running? The solution: a special muscle allows the giraffe to curl its tail tightly over its back when running. Like many other animals, the giraffe is a smorgasbord of features – each one specifically designed for its benefit. Does evolution explain any of this? Not really. No step-at-a-time process explains the change from horse-like creature to the giraffe. God is simply the Master Designer.

**Whence then cometh wisdom? and where is the place of understanding? God understandeth the way thereof, and he knoweth the place thereof.**
**~ Job 28:20,23**

The process of making light from chemicals is called bioluminescence. The most common example is the light produced by fireflies. But there are dozens of vastly different creatures which have this ability. From single-celled organisms called dinoflagellates to glow worms found in caves; from deep-sea fish to googly-eyed glass squids; there is a vast array of creatures with an ability to mix varying forms of luciferin and luciferase to produce light at will.

It turns out that each of these creatures uses a slightly different variation of the key chemicals to produce light. One would think that closely-related organisms should have similar luciferins and luciferases, while creatures farther apart on the evolutionary sequence would have much different versions of such chemicals. NO SUCH PATTERN EXISTS. Thus according to those who have extensively studied this subject, "bioluminescence is estimated to have evolved independently at least 40 times."[1]

It is difficult to believe that even one creature happened upon the ability to produce light by producing and mixing two complex



chemicals. But to believe it happened independently 40 different times is beyond credibility. Furthermore, these chance happenings far exceed mankind's ability to reproduce the same results. A firefly's luminescence is 88% efficient, while the light produced by the best luminescence reaction developed by mankind is a mere 23% effecient.

***O taste and see that the Lord is good:***
***blessed is the man that trusteth in him.***
***~ Psalm 34:8***

How did humans acquire language? Evolutionists believe it developed from primative people grunting and hooting. But



the astonishing rise of a new language by deaf children gives us insights into language development.

In 1980, in Nicaragua, 500 deaf children came together for the first time in a new school. Up until that time, these children had lived scattered in all parts of the country, communicating with hearing relatives via gestures. There was no established form of sign language; each child had his own set of gestures with little in common. But when they came together in the school, they quickly developed their own unique language, a completely new form of sign language.

At first this language was very rudimentary, but before long it became a fully developed language with rules of grammar and syntax. Rutgers' behavior neuroscientist, Judy Kegl, described it as "the first documented case of the birth of a language." Little kids, three or four years old, generated a full-fledged language without outside help! Their language was entirely of their own making. There is nothing that they could have used as a model. It was clear evidence that we are preprogrammed for language!

Adam and Eve were the first humans, and they had language to communicate with each other and God. Even though humans use vastly different languages, they seem to have an innate ability to produce complex communication systems. God programmed our brains to communicate using complex languages. No animal comes close to possessing this ability. Humans are not mere animals!

***...let us go down, and there confound their language, that they may not understand one another's speech...***

***~ Genesis 11:7***

The biggest problem for evolution is still the sudden appearance of every basic body structure – suddenly and without transition[1]. This fact of science clearly places the theory of evolution in the category of faith…rather than science. Furthermore, it is a faith that flies in the face of the evidence rather than acknowledging the evidence.

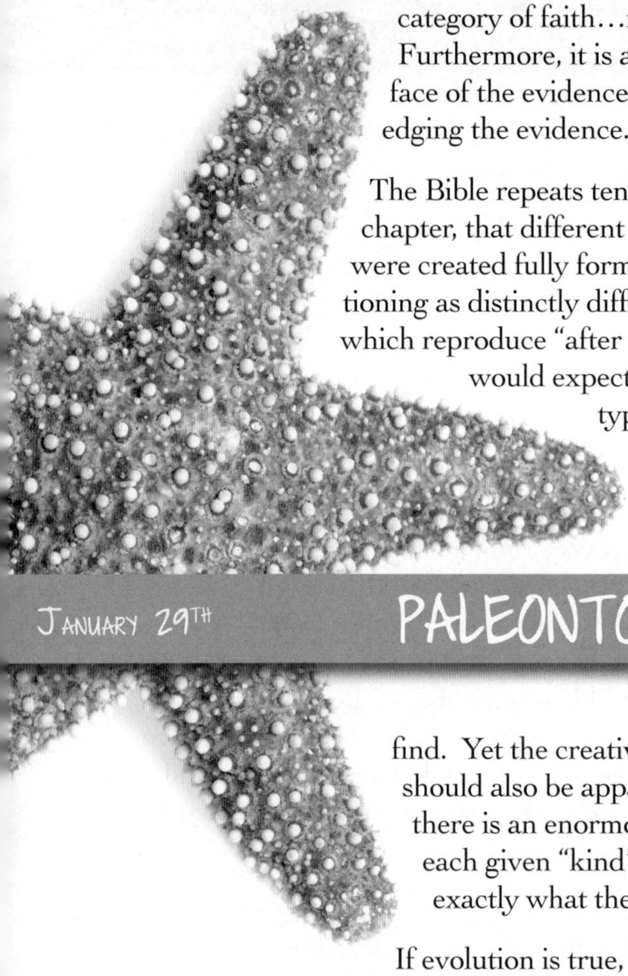

The Bible repeats ten times, in the first chapter, that different kinds of creatures were created fully formed and fully functioning as distinctly different creatures – which reproduce "after their kind". Thus, we would expect distinctly different types of creatures to be found in the rock layers of our planet – exactly what we find. Yet the creativity of the designer should also be apparent because today there is an enormous variety seen within each given "kind" of creature. This is exactly what the fossil record shows.

If evolution is true, some simple single-celled form of life must have somehow slowly been transformed into every other form of life. How does one go from a bacterium to coral and leave no evidence of the transformation? Where are all the in-between creatures? How do you slowly transform a single cell to a starfish…sponge…clam…trilobite…nautiloid…seaweed…fish with backbone…ALL without leaving ANY transition?????

**There is a way which seemeth right unto a man, but the end thereof are the ways of death… The simple believeth every word: but the prudent man looketh well to his going.**

**~ Proverbs 14:12,15**

There is no place on Earth which is not overflowing with life:

- **The frozen polar caps** have life as majestic as the lumbering polar bear, comical as the waddling penguin and persevering as the Arctic Tern. Each is specifically designed to survive within this harsh frigid environment.
- **Arid deserts** come alive at night with lizards, snakes, scorpions, and a dizzying array of creatures with features designed to conserve moisture.
- **Deep ocean vents** spouting toxic, boiling, mineral-laden fluids are surrounded by microorganisms designed to thrive in this harsh environment.

## INSPIRED BY BIOLOGY

JANUARY 30TH

Contrast this with what we know about every other place in the universe:

- **The moon** is a barren, airless, crater-scarred wasteland…totally devoid of life.
- **Mercury** is a furnace hot enough to melt lead…totally devoid of organisms.
- **Venus**, Earth's "sister planet", is a caldron of toxic gases…devoid of life.
- **Mars** is a desert planet with no indication of life.
- **All outer planets** are frozen gas giants completely devoid of the beauty, sound, movement, and purpose provided by plants and animals.
- **Planets detected around distant stars** are too large, too close, or too far from their stars to support life. Essentially, all known stars have characteristics incompatible with life.

So why do we hear so much about finding life on other planets or galaxies? Because if life developed by chance, unguided processes here on Earth, then life must have also developed without God elsewhere in the universe. Yet only faith in the impossible allows belief that life exists elsewhere in the universe. Earth is an unparalleled paradise – perfectly designed for the existence of life. Our planet is specifically for our benefit, by God!

**And God said, "Let the waters bring forth abundantly the moving creature that hath life…"**

**~ Genesis 1:20**

Have you heard that your eye is wired backwards? Yes, the wiring faces toward the incoming light, while the photoreceptors face the back of the retina. When light enters the eye and falls on the retina (back of the eye), it has to go through 10 layers to the very back of the retina for the photocells to be activated. Why would an all-powerful, all-knowing, perfect Designer invent such a convoluted system? Atheists cite this as "evidence" that God does not exist.



But let's examine the evidence.

- Photocells are highly active and need a continual feeding of blood rich nutrients. This blood-rich lake lies behind the photoreceptors. If the photoreceptors faced forward, this lake of blood would be in front of the photocells, and we would be looking through lots and lots of blood.
- The photocells are one of the most active areas in our bodies and need to be replenished every 7 days. What to do with old burned out debris? Enter the macrophages (like microscopic garbage collectors). These garbage collectors operate within the pooled blood and if they were in front of the photoreceptors, it would again be impossible to see clearly.
- So how does the light get to the backward facing receptors? Muller cells in the front of the retina collect and funnel the light to the photoreceptors. Muller cells work like fiber optics – only better – because they cover the entire surface of the retina collecting and funneling the maximum amount of light to the photoreceptors.

What appears at first glance to be a "bad design" is in reality a superb "God design". The backwards wiring of the eye allows a rich supply of blood for the receptors, while the Muller cells with their "fiber optics" move light to the photoreceptors. Could this have all happened by chance mutational changes from some preexisting light sensitive patch? Not a chance!

***O Lord, how great are thy works! and thy thoughts are very deep.***

***~ Psalm 92:5***

# FEBRUARY

*He is no fool who gives what he cannot keep*
*to gain what he cannot lose.*

**- Jim Elliot** (1927-1956)
martyred missionary to Ecuador

*Greater love hath no man than this,*
*that a man lay down his life for his friends.*

**- John 15:13**

God is a God of infinite creativity and variety. This is the God who tossed into existence 1,000,000,000,000,000,000,000 stars and keeps them whirling as they plunge through space at unimaginable speeds. And yet with mathematical precision, we can know when a



comet will visit years in advance or know when an eclipse will take place. Even the ancient Mayans accurately calculated the position of planets thousands of years into the future.

This is the same God who selected a single planet and filled it with a bewildering number of life forms. On the sixth day, God made his final preparations for man. Scientists have classified millions of these creatures. When we tally all the creatures that God made on day 5 & 6, we find there are at least 800,000 different kinds of insects, 30,000 kinds of fish, 9,000 kinds of birds, 6,000 kinds of reptiles, 3,000 kinds of amphibians and 5,000 kinds of mammals. God is truly a God of variety! Evolution may imagine life struggling to emerge and finally succeeding at a lonely, isolated place. But these mutations over time simply cannot explain life's incomprehensible variety.

This God of infinite power, creativity, and wisdom chose to enter into human history in order to bring rebellious humans back into fellowship with Himself. If Genesis 1 was a psalm, it would almost certainly finish with a resounding "Selah", meaning: "There, what do you think of that!"

**And God said, "Let the earth bring forth the living creature after his kind, cattle, and creeping thing, and beast of the earth after his kind:" and it was so. ~ Genesis 1:24**

Here are just a few of the properties of air – which make life possible upon the Earth:

- **Oxygen for breathing:** The amount of oxygen in the air is 21%, which is just perfect. If there were less, then the animals would be breathless. If there was too much, fires would burn continually.
- **Carbon dioxide for photosynthesis:** Plants convert the carbon dioxide into other carbon compounds.
- **Nitrogen for plants:** Nitrogen is needed for cellular growth.
- **Radiation screen:** Air acts like a screen preventing the sun's harmful ultraviolet radiation from coming through.
- **Medium for color:** We have the beauty of skies because of air. Compare this to our moon with no air, causing a dark sky.
- **Medium for sound:** Our air makes it easy for sound to travel. Air molecules vibrate in the air so that we hear birds singing and insects chirping.
- **Medium for vision:** Air is colorless, allowing clear vision.
- **Medium for smell:** Air is odorless, therefore, we are able to smell delicate flowers.
- **Medium for thermal insulation:** Air acts like a thermal blanket for Earth, preventing temperature extremes. On the moon, where there is no air, the temperature when the sun is shining is 250°F (123°C). When there is no sun, the temperature drops to -380°F (-231°C).
- **Medium for pressure:** Air has pressure that keeps the different parts of our bodies in place. Astronauts' faces in space become very puffy because of lack of air pressure.

## INSPIRED BY DESIGN

February 2nd

To design a <u>single substance</u> to perform so many complex functions represents a masterpiece of design! Engineers often design products to optimize a single functional purpose. Yet, air has many wonderful properties – providing powerful evidence of God's unfathomable intellect and creativity.

**Blessed be the Lord God, the God of Israel, who only doeth wondrous things!**

**~ Psalm 72:18**

The Arctic woolly-bear caterpillar is known as the longest living caterpillar because it lives for 14 years as a caterpillar. That means it goes through 13 seasons of being completely frozen through the winter and thawing out each spring before becoming a moth in its 14th year (most caterpillars live only a few weeks before turning into moths). Late in summer, the Arctic woolly-bear caterpillar starts to make glycerol, a type of antifreeze. Normally when fluid freezes, the expanding crystals rupture the cell's membrane, allowing the fluid to leak out. The caterpillar uses glycerol to control the process of crystal formation. As the temperature



drops, the wooly-bear caterpillar's gut freezes first, then its blood, and then the caterpillar shows no heartbeat or breathing. It is completely frozen. As spring arrives, it thaws out, ready to eat and grow again.

How does evolution explain this? How many attempts to survive the winter did the caterpillar try? If it failed, the result was death – an evolutionary dead-end. When did a certain caterpillar "get it right" and survive? Remember, there had to be two, a male and female, surviving to produce eggs and continue the species. The original Arctic woolly-bear caterpillar had to make this antifreeze so that its cells would freeze without rupturing not just once for one winter, but 13 times for 13 winter freezes, always remembering to produce the antifreeze right before winter arrived. Then it had to learn to completely rearrange its body structure to turn into a moth on the 14th spring. How do evolutionists explain this? They don't! The Arctic woolly-bear caterpillar was designed to survive many winters. When you see design, there must be a Designer, and that Designer is God.

**Where is the wise? where is the scribe? where is the disputer of this world? hath not God made foolish the wisdom of this world? For after that in the wisdom of God the world by wisdom knew not God**

**~1 Corinthians 1:20,21**

Every day, in cow pastures around the world, shots are fired.

Most people have never seen these shots because they are fired by a ½ inch tall fungus called Pilobolus. This fungus lives on the fresh dung of horses, deer, cattle and other grass-eaters. Like all other fungi, they decompose dead or decaying matter. They start as spores that grow into thin threads feeding on the manure. Soon stalks begin to sprout and grow. Near the top of the stalk, a bulge develops like a cannon barrel. On the top of these stalks appear the cannonball-like spores. In the morning sun, this bulge fills with water. When the water pressure gets high enough, the bulge bursts, launching the spores up to six feet away.

Pilobolus spores need to be a good distance from the manure, so the cows will eat them. Cows find their own dung repugnant and won't even eat the adjacent grass. The far flung spores will land on a blade of clean grass, the grass is eaten by the cow, the spores pass right on through the guts of the cow and end up deposited in a fresh pile of dung…thus beginning the life cycle of the Pilobolus again.

What if the spores were not shot far away? They would not be eaten. What if there was no protective coating on the spores when they were being processed through the cow's digestive system? They would be destroyed. What if the fungus was not there to turn the cow dung into soil? We could be knee deep in cow dung! Each step and component of this system is dependent on the other parts. This is an irreducibly complex system.



We see God's amazing ingenuity in solving something as mundane as manure decomposing. If you listen really closely you may even sense the Holy Spirit saying, "Just wait until they figure out how I solved this problem!"

**He causeth the grass to grow for the cattle, and herb for the service of man: that he may bring forth food out of the earth.**

**~ Psalm 104:14**

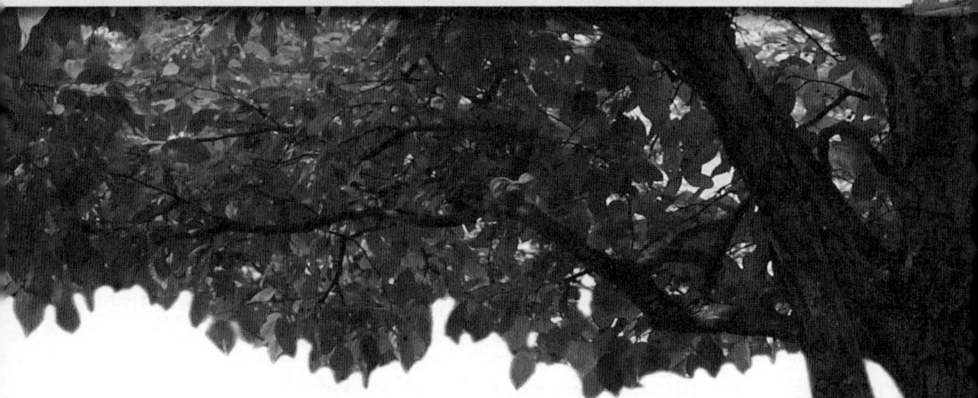

Over a course of a single growing season, forests pump millions of gallons of water into the air. This is why walking through a forest on a hot summer day is so refreshing. Even a single tree pumps dozens of gallons of water from its roots to its leaves each day. How is this possible? The hot sun on the leaves causes water to evaporate from the leaves. This evaporating water creates a vacuum within the leaves' cells. Because water molecules like to "stick together", this vacuum pulls the water far up into a tree – almost like it is being sucked up a drinking straw. Water molecules have been tracked moving as fast as 25 miles per hour up through the trunk of a tree! Over 90% of the water which moves from the ground into the leaves of a tree is evaporated from the surface of the leaves. Only 2% of the water is used for photosynthesis. It is this massive amount of evaporation that provides the driving force to suck so much water out of the ground and carry it hundreds of feet into the air.

Even a small grove of trees removes tons of water from the soil and releases it into the air each day. It has been estimated that an apple tree gives off 16,000 pounds of water to the air during one growing season. This is an amazing system involving capillary flow, hydrogen bonding of water molecules, osmosis, and vacuum pressure. Trees are God's method of humidifying the world. And all this happens in total silence! What an example of God's engineering excellence.

**The trees of the LORD are full of sap; the cedars of Lebanon, which he hath planted...**

**~ Psalm 104:16**

How can we see distant stars in a "young" universe? This question is frequently asked of creation scientists. Many suggestions have been offered[1]:

- **Decaying light speed:** If the speed of light were faster in the past, early starlight could have moved more rapidly than today.
- **Relativist cosmology:** Time itself varies for objects in the vicinity of extreme mass or high speed. During Earth's formation a short time may have passed on Earth while a longer time passed in space, providing time for the light from faraway stars to reach us.

## INSPIRED BY COSMOLOGY

FEBRUARY 6TH

- **Mature Creation:** During the creation week, everything was made fully mature; trees bore fruit, and Adam was an adult. The universe was formed fully functioning with stars visible on Earth from the moment of creation on day four.

Big Bang evolutionists have a bigger problem, which they have never answered. Stars do not last for the assumed age of the universe (14 billion years) because they use up their energy supply. Therefore, cosmic evolutionists are forced to conclude that new stars are forming. Yet, there is not a single example of a new star "turning on" in the night sky. This is a devastating blow to their belief system, far more troubling than explaining how light could reach the Earth from distant stars in a recent creation.

***Knowest thou the ordinances of heaven? canst thou set the dominion thereof in the earth?***
***~ Job 38:33***

Not all eggs are the same; they come in a variety of sizes, colors and shapes. Robins have oval eggs that fit nicely in their dish-shaped nest. Screech owls have round-shaped eggs that lie at the bottom of a hole in a tree. Birds like the Killdeer have eggs that are sharply pointed allowing them to pivot on their small end. This is the perfect design for this



bird because the Killdeer's nest is on bare ground, and round eggs could roll away. The common Murre nests on a sea cliff ledge and lays their eggs on bare rock ledges. The egg's oblong shape prevents it from rolling off the ledge because when pushed or nudged, it can only roll in a circle. Eggs that are hidden away are usually pale or a solid color, while eggs laid out in the open are spotted, blotchy, and camouflaged.

There are more than 9,000 species of birds in the world. Think of all the variety of eggs! Eggs are not just "happy coincidences", but they reveal the hand of the Creator that designed their perfect shapes, sizes, and colors to fit into their nurseries – details preprogrammed into the creation to allow birds to fill all sorts of different environmental niches.

**For thou art great, and doest wondrous things: thou art God alone.**
**~ Psalm 86:10**

The pyramids of Egypt reveal ancient man's ingenuity. The Giza pyramid was built with 115,000 casing stones of polished



white marble, each of them weighing from 16-20 tons. A massive earthquake in A.D. 1301 loosened many, which were then removed to build Cairo. The joint thickness between these marble stones was a minuscule 0.020 inch. Inside the narrow gap was a type of cement which bonded the marble with greater strength than that of the marble itself. The composition of this cement is still a mystery to us. The largest blocks measured five feet high by 12 feet long by 8 feet deep, with an average variation of only 0.01 inch. This degree of exactness is diffcult even today.

When we examine any Egyptian pyramid we find that the four sides are equal in length within 0.1%. Modern builders do well with 1-2%. The pyramids were also built on absolutely level bases with no sagging. It has been estimated that it would take 100,000 men to construct a typical pyramid and they would need to set a new block every 20 minutes for the pyramid to be completed in 20 years. There are over two million quarried blocks in the Great Pyramid.

After the Flood and the tower of Babel mankind migrated across the globe and spread their technological knowledge. For instance, some of Noah's relatives settled in Egypt. The first pyramids of ancient Egypt were the best, with the later ones becoming smaller and shoddier. Almost every ancient culture (all of which spang into existence shortly after the flood) showed signs of great technological knowledge and creativity. Ancient man was not some primitive cave man but was made in God's image and highly intelligent.

**And as for Zillah, she also bore Tubal-Cain,**
**an instructor of every craftsman in bronze and iron.**
**~ Genesis 4:22**

# February 9th — BIBLICAL ACCURACY

INSPIRED BY

In the Gospel of Mark, there is a fascinating account of Jesus healing in a two-step process. First, Jesus spat on the blind man's eyes, restoring partial sight. Then Jesus put his hands on the blind man's eyes, giving him totally clear vision. Between the two steps, the man saw "men as trees, walking". Today we call this agnosia, a condition where the brain does not have the ability to interpret what is being seen. There are many case histories of formerly blind people recieving sight but not perceiving what they are seeing.

For example, Virgil, who had been blind from childhood, had his eyesight restored. When the bandages were removed, Virgil had no idea what he was seeing. His brain could not make sense of the images that his optic nerve was sending to his brain. He could see the parts of a cat clearly (such as the tail and the nose), but the cat as a whole was blurry. "Trees didn't look like anything on Earth," he said a few days after his operation. It was a month later before he finally saw the trees the way we do. Blind people who regain their eyesight require time to develop new pathways in the brain's visual cortex before perceive what they are seeing.

When Jesus healed the blind man, it was a two-step process. First, he healed the blindness in the eyes. Second, he healed the neurological pathways and connections in the brain. Why did Jesus do it this way? We don't know for sure, but it does give a stamp of authority to the authenticity of the account. Back in Jesus' day, there were no surgical ways of correcting blindness from birth. The writer could not have known about the problem of agnosia in the newly sighted. Perhaps Jesus wanted us (those who live in a time when such knowledge is available) to recognize that Scripture is not only medically accurate in its observation of historical events but miraculous in nature. Jesus did perform miracles.

**And he took the blind man by the hand, and led him out of the town; and when he had spit on his eyes, and put his hands upon him, he asked him if he saw ought. And he looked up, and said, "I see men as trees, walking." After that he put his hands again upon his eyes, and made him look up: and he was restored, and saw every man clearly. ~Mark 8:23-25**

Evolutionists are in trouble from the very beginning. The Big Bang has to explain where everything came from before the universe came into existence. Cosmic evolutionists are left with only two possibilities. Either all of the matter and energy in the universe have been around for an infinite time, or everything just made itself out of nothing. Neither work because both contradict the most firmly established laws of science.

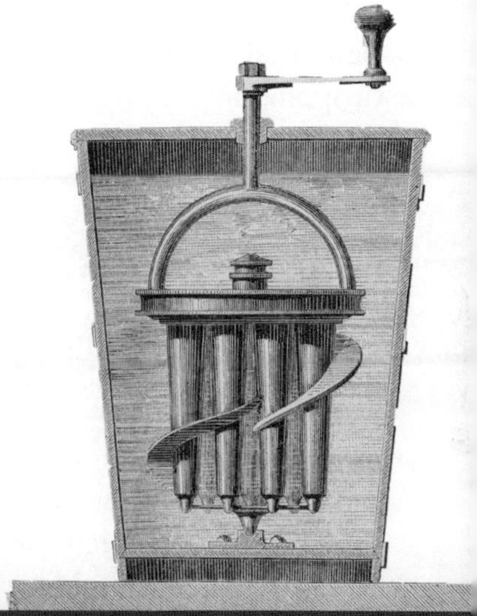

The first law of science shows that matter and energy can be neither created nor destroyed. Therefore, based on every observation, experiment, and theory ever tested… the universe could not have made itself. In spite of this, it is commonly taught that nothing somehow turned into the "cosmic egg" that explosively expanded into our current universe. But nothing ever turns into something. Nothing

## INSPIRED BY PHYSICS — FEBRUARY 10TH

can do… nothing. Nothing can only become nothing. Nothing is the absence of everything: no energy, no matter, no space. Also, if the universe were infinitely old, everything would be randomly dispersed and uniform in temperature. However, it isn't – so the universe cannot possibly be infinitely old. It must have had a relatively recent beginning.

The most tested, foundational laws of science clearly show the impossibility of the Big Bang as an explanation for our existence. Yet these same laws of science are perfectly compatible with the recent creation of the universe by God.

**By the word of the Lord were the heavens made; and all the host of them by the breath of his mouth.**
**~ Psalm 33:6 (NKJV)**

For decades it was believed that the praying mantis had no sense of hearing because there are no hearing membranes on its head. However, in the late 1980's, neuroscientist David Yager discovered that most mantids (praying mantis species) have a single hearing membrane buried deep in the center of their thorax (chest area).

Dr. Yager has shown that flying mantids can detect the ultrasonic chirps of an approaching bat and take evasive action much like a fighter jet – pulling up into a stall position, turning sideways, and dropping into a powered dive toward the ground.



This complex aerial maneuver usually succeeds in losing the trailing bat.

Not only does no other insect possess such an ultra sensitive "ear", but there is no known ancestor or evolutionary path explaining its existence. The evidence indicates that this one specific creature was designed with this amazing ability. Further frustrating any supposed evolutionary explanation, fossil mantids had been shown to possess the same features long before bats were supposed to have evolved, making the need for such a feature unnecessary. The Bible clearly teaches the creation of every different "kind" of creatures at essentially the same time – making such features no surprise.

So how do evolutionists explain the existence of a unique form of hearing in a creature long before such an ability would be useful? As usual, they fall back on faith disguised as science. Yager and his colleagues "speculate a different [unknown] kind of insectivorous predator could have chased them during the Cretaceous Period." This is storytelling, not science.

**For God speaketh once, yea twice, yet man perceiveth it not.**

**~ Job 33:14**

Charles Darwin believed that if evolution were true it would be found in the fossil record. He believed that the fossil record would show a species gradually changing from simple to greater complexity. But over the last hundred years, these transitional forms have simply not been found. What has been found are fully formed, fully functional, very different forms of life. Evolutionists are especially confounded by a layer close to the bottom of the geological column (that sediment buried first or deepest during Noah's Flood). What we find in this rock layer is a veritable explosion of life:

"The Cambrian 'explosion' … is one of the biggest mysteries of biology. It is at this point in the fossil record that a multitude of animal forms suddenly appears, for reasons that are not well understood.[1]"

# INSPIRED BY PALEONTOLOGY

February 12th

Within this particular rock layer we find every basic phylum (body type) of life on earth. This sudden appearance "out of nowhere" poses significant problems for evolution. This layer does not show animals evolving from simple, single-celled organisms into more complex structures. The Cambrian explosion shows the sudden burst of great variety. Evolution from one type of creature into a completely different type of creature has never been observed in the natural world or the fossil record. Remember, science is based on what we can observe.

*When wisdom entereth into thine heart...understanding shall keep thee:*

*~ Proverbs 2:10,11*

Water makes our planet unique. When astronomers are looking for life on other planets, they look for water. One of the wonderful properties of water is its ability to hold heat. In fact, our oceans and seas act like big thermostats that moderate the Earth's weather. If you take a water balloon and hold it directly over a candle flame, it will not break. The water in the balloon absorbs the heat. Water's ability to store heat energy (called the heat capacity) is one of the highest of any known liquid or solid in nature.

During the day, the Earth's seas and lakes soak up the sun's heat; as a result, the Earth stays fairly cool. During the night, heat is released into the atmosphere, keeping the surface from freezing. If our Earth did not have this tremendous amount of water, our temperatures would vary greatly. We can easily see the ocean's moderating effect on temperature by comparing Seattle, Washington with Bemidji, Minnesota. Both cities are at the same latitude, 47 degrees north. However, Bemidji averages 27 degrees colder in the winter than Seattle. Seattle's temperature is moderated by the Pacific Ocean water.

Consider the Gulf Stream; it picks up great quantities of heat from the equator and moves northward to Western Europe and the Arctic region. The Gulf Stream carries the equivalent of 1,000 Mississippi Rivers in volume, bringing with it the warm waters of the equator. Because of this water, Northern Norway's average temperature is 75 degrees warmer than other inland locations of this latitude. The Gulf Stream's warmth can even melt a 150,000 ton iceberg in a single week.



God cares about our life on Earth. The unique heat-holding property of water moderates the Earth's temperature. This did not happen by accident or chance. Even the heat-holding characteristic of water was designed by God for our benefit.

**Who hath measured the waters in the hollow of his hand?...**
**Who hath directed the Spirit of the Lord, or being his**
**counsellor hath taught him?**
**~ Isaiah 40:12,13**

What is the opposite of hot? Most people would say cold, but you cannot make something colder by adding more "cold". You can add and remove heat from an object, but not cold. There is no theoretical limit to how hot something can become, but once all heat is removed it cannot get any colder. Cold is actually the absence of heat.

What is the opposite of light? Most people would say dark, but you cannot make a room darker by adding more and more "dark". You must remove light. Once you remove all light, such as in a deep cave, it becomes absolute darkness and cannot get any darker. There is no theoretical limit to how bright something can be, but darkness is the absence of light.

What is the opposite of God? Most people would say the devil. Many popular movies, such as *Oh, God!*, promote this idea, but the devil is no comparison to God. Others would say evil is the opposite of God, but evil is actually like darkness and cold; it does not exist because one person has more of this quality than another – evil is the absence of love. The less love we have, the more evil we become. So where does love come from? God. GOD IS LOVE.

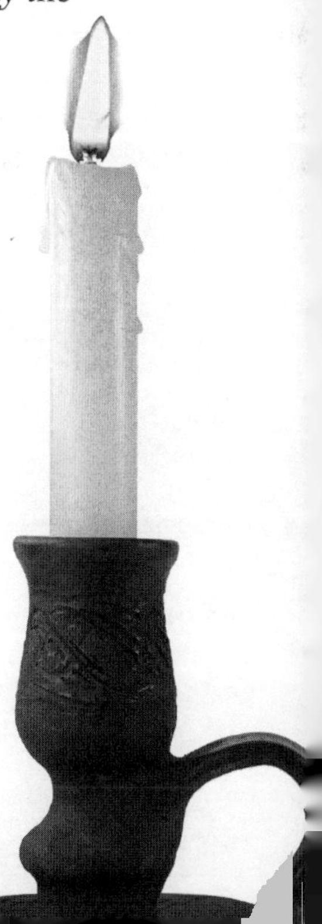

It is interesting that Jesus constantly compared Himself and God to love and light. If we want more of these things in our lives, we must draw closer to the source of both.

**I am the light of the world: he that followeth me shall not walk in darkness, but shall have the light of life.**

**~ John 8:12**

Humans have around 50 facial muscles, of which half are dedicated to making expressions. In contrast, apes have fewer than 30 facial muscles, of which zero are dedicated to making expressions. Apes have coarse muscles in their cheeks, while humans have very delicate cheek muscles. The delicate cheek muscles allow humans to make many types of smiles. In contrast, apes' cheek muscles are not able to produce smiles. There are many different types of



expressions such as smiling, laughing, anger, pain, surprise, and boredom. Each of these expressions, and many more, have varying degrees of intensity. Researchers have made the amazing claim that humans are able to make up to 10,000 different expressions.

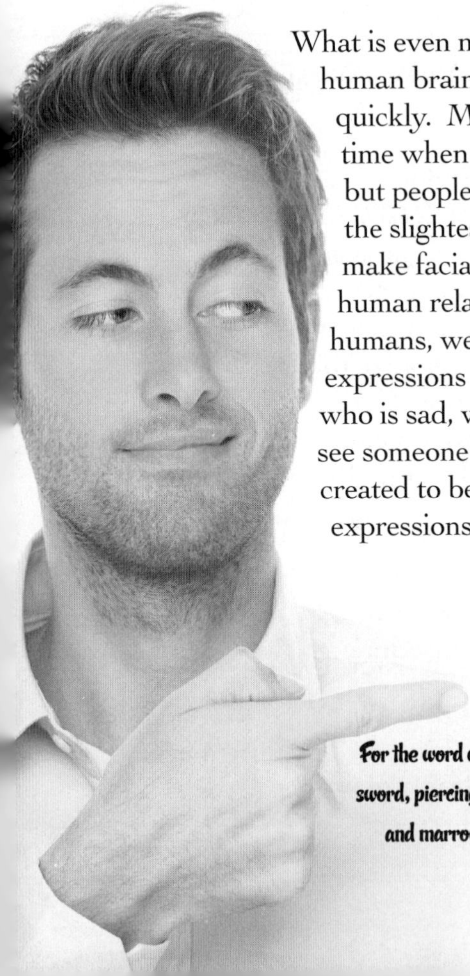

What is even more surprising is the ability of the human brain to discern the subtle differences quickly. Most people have experienced a time when they tried to hide their emotions, but people are very perceptive in detecting the slightest facial expression. The ability to make facial expressions is very important in human relationships and communication. As humans, we are constantly observing the facial expressions of others. When we see someone who is sad, we ask what is wrong. When we see someone smiling, we smile back. We were created to be emotional beings, and our facial expressions reveal that design.

**For the word of God is quick, and powerful, and sharper than any two-edged sword, piercing even to the dividing asunder of soul and spirit, and of the joints and marrow, and is a discerner of the thoughts and intents of the heart.**

**-Hebrews 4:12**

Humans have the most expressive, communicative faces of any organism ever created. The human face is controlled by 53 different muscles which enable us to express an incredibly wide variety of messages and emotions. It has been said that 90% of human communication is non-verbal (facial expressions, tones of the voice, body postures, etc.). How much more difficult this would be if the human face were not made to be so expressive.[1]

## INSPIRED BY ANATOMY

FEBRUARY 16TH

Charles Darwin made numerous blunders while promoting the viewpoint that humanity ascended from previous forms of animal life. In order to promote his fallacious viewpoint, Darwin published a book in 1872 entitled *The Expressions of the Emotions of Man and Animals* in which he altered and doctored photographs to make the expressions of man look more animal-like. He did this in order to promote the concept that animal and human expressions of emotion are similar. This concept has since been thoroughly disproven.[2]

Surprise causes an identical facial expression across every human culture in the world. In every culture, when a person is surprised, his brow goes up and his eyes widen. It is almost as if the body is saying, "What I had believed to be true has been shown to be wrong. I'd better open my eyes wider, take in more information, pay attention, and re-evaluate what I believe to be true." Good orators and storytellers have always understood that the element of surprise is an important tool in gaining acceptance of ideas. Just presenting facts and figures seldom changes perceptions, and our very facial response reflects this reality. In contrast, anger is characterized by a narrowing of our eyes such that we can intently focus on a specific, potential threat. Even the muscles of our face and the involuntary expressions in response to unexpected happenings are designed with intricate care for our benefit.

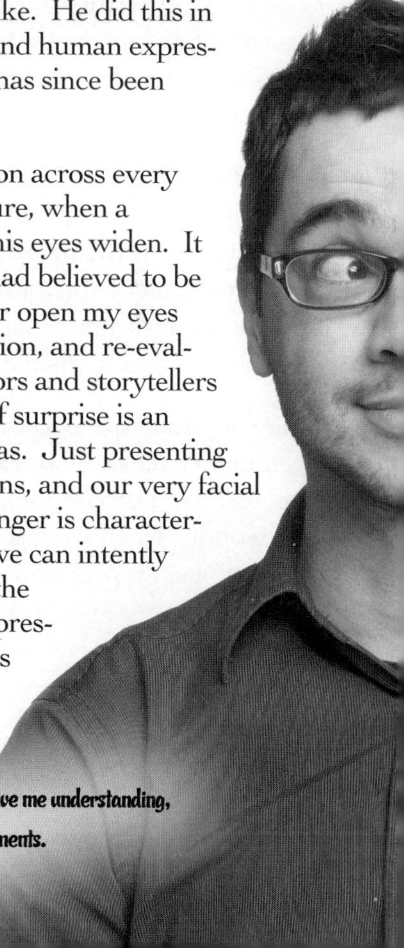

**Thy hands have made me and fashioned me: give me understanding, that I may learn thy commandments.**
**~Psalm 119:73**

Leaf cutter ants are mushroom farmers. They cut leaves and haul them back to their underground colony. Other worker ants reduce the leaves to mulch. This mulch is the compost on which the fungus grows. Leaf cutter ants use this fungus for food. However, this fungus can be destroyed by mold.

Leaf cutter ants are also pharmacists. The worker ants that are tending this fungus farm use a special antibiotic to kill the mold. This antibiotic, actinomycetes, is carried on the ants' bodies and scattered about as they tend the farm.



Leaf cutter ants are also treasure hunters. Because this antibiotic is needed, other ants go out and look for it in the soil and bring it back. They have to find the right antibiotic in the right soil to protect their fungus and then walk around the farm spreading the antibiotic.

How did leaf cutting ants develop their farming techniques? How did they know how to cut and compost the leaves in order to make a mulch to feed the fungus? How did they know that a certain soil contained exactly the antibiotic that would kill that specific mold? How did they know to have the worker ants carry this antibiotic on their bodies and to drop it off as they tended to the farm?

They did not have to know, nor did all this develop by some seemingly magical "natural selection" or other evolutionary process. God brought it all together for the great pleasure of seeing our wonder when we discovered this and other marvels He has created. He preprogrammed these many characteristics into the ant's DNA.

*Blessed be the name of God forever and ever, for wisdom and might are his. ~ Daniel 2:20*

It is not just the number of changes required to change a single-celled organism into a human that is mindboggling, but the order in which these changes have to happen. It does not aid an organism to have a leg until there is a nervous system to control the leg. An eye is useless until the brain has developed the ability to collect and interpret the information from the eye. So why develop this ability until the eye exists?

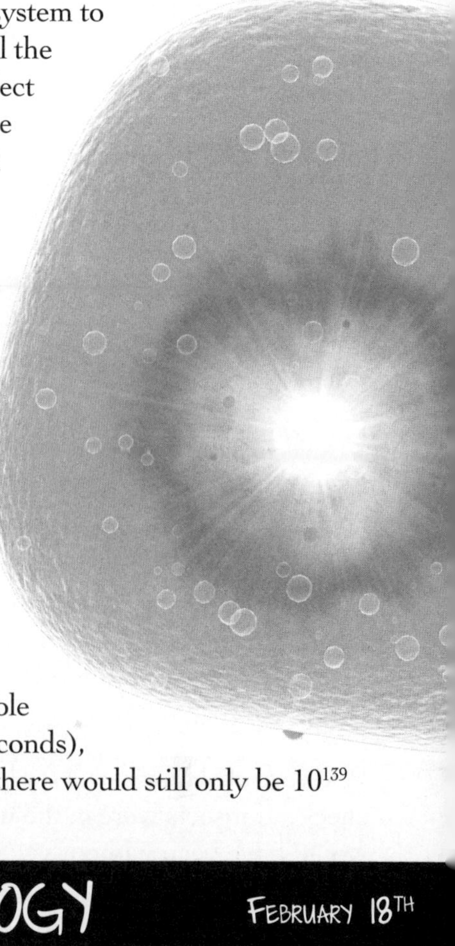

Even if only 1000 beneficial mutations are needed to change one creature into a "slightly" more advanced creature, the mutations would still have to occur IN THE RIGHT ORDER. The odds of this happening are like flipping a coin and getting heads 1000 times in a row. This would happen once in $10^{301}$ attempts. Even if every subatomic particle in the entire known universe ($10^{80}$ particles) mutated at the fastest possible rate (Planck time = once evey $10^{-42}$ seconds), and had done so for 15 billion years, there would still only be $10^{139}$ mutational possibilities.

## INSPIRED BY MICROBIOLOGY

FEBRUARY 18TH

There is simply nowhere near enough mutations or time available to explain the transformation of one form of life into another. And this is a mere 1000 changes! The transformation from amoeba to man would require millions of changes – all in the right order. Based on what we know of time, space, matter, and probability, evolution is an absolute impossibility

**How precious also are thy thoughts unto me, O God! How great is the sum of them! If I should count them, they are more in number than the sand.**

**~ Psalm 139:17,18**

The yellow throated warbler is a beautiful songbird that has been observed to eat 70 aphids per minute. When the warbler had finished eating – 40 minutes later – he had eaten almost 3000 aphids. A nighthawk can eat 500 mosquitoes in one feeding. It has been estimated that just the chickadees in Michigan consume over 8 billion insects every year.

Evolutionists say that insects appeared millions of years before birds evolved. If birds were not here to eat insects, insects would have filled the world and destroyed all its plants.



For example:

- One pair of Colorado potato beetles, if left unchecked for one season, will produce over 60 million offspring.
- One female fly, if left unchecked for the summer, will have 143,875 bushel baskets of offspring.
- Aphids can produce 13 generations during a single season. If each mating produced over 100 females, at the end of this process, there could be ten sextillion aphids.

Birds are absolutely vital for keeping the insect population in check. If insects were in the world long before birds, our planet would have been without vegetation and very desolate. Insects need to be kept in check; birds do that. Insect population growth testifies to what we read in Genesis 1; God created everything in one week, including the birds and the insects.

**O sing unto the Lord a new song; for he hath done marvellous things...**

**~ Psalm 98:1**

On a clear night one can see about 3,000 stars, yet in Jeremiah 33:22 it states, "As the host of heaven cannot be numbered, neither the sands of the sea measured." Jeremiah wrote that no one would be able to count the stars, contrary to the scientific knowledge of that time. Not until the 17th century did we get a glimpse into the vastness of our universe when Galileo introduced his new telescope. But only when the 20th century Hubble telescope focused on one area with time-lapse photography did we begin to realize just how great the number of stars really is.

Today astronomers have estimated the number of stars to be ten billion trillion, which is a 1 followed by 22 zeros! No human can



live long enough to count all the stars. If a computer were used to count the stars (the fastest computers can preform about 10,000 million calculations in one second), it would still require millions of years of non-stop counting to count all the stars! Even though in Jeremiah's time people **could** number the stars, God knew "they cannot be numbered". God told us this through his prophet Jeremiah in 600 BC! The Hubble telescope, some 2,600 years later, confirmed that the stars really are countless, just like the grains of sand on the seashore.

Wherever the Bible speaks on science, we eventually find out it is true. What a great and powerful God we have!

**He telleth the number of the stars; he calleth them all by their names.**

**~Psalm 147:4,5**

Imagine a bird feeder. Where did it come from? Of course you would say someone made it, but how do you know? Did you see someone making it? How do you know the bird feeder has a maker? Even though you did not see someone making the feeder, you can see some organization and complexity. Even though a feeder is a simple structure, it still reveals design that could not have come about by accidental changes or chance over time. Someone must have made it. Why did they make it? Perhaps so they could enjoy watching birds!

The existence of a bird feeder leads to other questions: Can the bird feeder fly? Can the bird feeder grow feathers? Can the bird feeder make new birds? NO to all three!

Yet the bird can do all of these things. Evolutionists say the bird had no creator; that it simply happened by chance. Is this logical? If the feeder (which is far less complex than a bird) had a designer, then so did the bird. When we see a bird feeder, we know there is



a bird feeder maker. When we see a bird, we know there is a bird maker. That "bird maker" is God.

**And out of the ground the LORD God formed... every fowl of the air...**
**~ Genesis 2:19**

It has often been stated that science is a "self-correcting discipline". Many people believe that a scientist who makes a new and signifi-



cant discovery receives instant fame and glory. In reality, the very opposite is true. Scientific thought becomes entrenched in dogma (not unlike religious doctrine), and those who oppose commonly accepted beliefs are all too frequently opposed and even vilified.

In 1982, after years of research, Dr. Barry Marshall presented a paper showing that bacteria cause peptic ulcers. At the time it was widely believed that ulcers were caused by stress and that bacteria could not survive in the acidic environment of the stomach. His research findings were ignored by essentially all other experts in the field. Despite clear evidence that he was correct, his work challenged the reigning paradigm in the field and was not even considered a possibility. Yet he knew that much pain and suffering could be alleviated if he could get other doctors to accept the evidence he had presented. Finally, in desperation, Dr. Marshall drank a vial of liquid containing the H pylori bacteria and gave himself peptic ulcers. Even then, it was ten more years before his work was widely accepted. Today it is acknowledged that up to 90% of ulcers are caused by bacteria. Twenty-five years after his discovery, Dr. Marshall was awarded the Nobel Prize in medicine.

If such a minor discovery requires dramatic action to wake up those opposed to the truth, imagine how hard it is to convince people trained to think in terms of billions of years of a recent creation. An ancient age for the Earth is THE cornerstone for evolutionary thought and explaining life without God. Huge periods of time are absolutely required for the belief in naturalism. Those caught in the paradigm of explaining everything by natural processes routinely ignore the evidence for a recent creation rather than accept the evidence at face value.

**Blessed is the man that walketh not in the counsel of the ungodly, nor standeth in the way of sinners, nor sitteth in the seat of the scornful.**

**~Psalm 1:1**

How do tiny flatworms called monogeneans testify to creation? They are parasites that live on the skin, fins, and gills of fish.

They first take hold of the fish with hooks. But in order to feed off the fish, they glue their mouths onto the fish with superglue. This flatworm has two glands within its head that make two non-sticky components needed to form the glue. These components are extruded through tiny holes beside the mouth. Only when the two components come together do they form a glue, just like a two-part epoxy resin. This glue is extremely strong and adheres to wet, slimy surfaces, even under water. The glue is delivered in a non-sticky form so that it does not glue the flatworm's mouth shut before it comes in contact with the fish. The glue cures quickly, it is stable and durable, and it can be dissolved when the flatworm needs to leave.

When researchers discovered this lowly form of life making superglue, they marveled at its amazing properties. How did this little flatworm develop such marvelous glue? How did it learn to keep the two components separated? How did it develop the delivery tubes? How did it learn to use the glue to its advantage? How did the worm learn to dissolve the glue? How could evolutionists believe this all happened by accident and chance?

**To him who alone doeth great wonders: for his mercy endureth for ever**

**~ Psalm 136:4**

Recently, a team of scientists wanted to test carbon-14 in coal and diamonds. Carbon-14 has a half life of 5,730 years. This means that in 5,730 years, half the radioactive carbon-14 decays away leaving behind non-radioactive material. Things that are said to be millions of years old could not have any carbon-14 present. Coal is supposedly hundreds of millions of years old.

Ten coal samples were tested; supposedly, they were 34 to 311 million years old. All ten coal

samples contained carbon-14. Then they tested diamonds. Diamonds are supposedly a billion years old. All of the diamonds tested contained carbon-14. Not a molecule of carbon-14 should have been left in either coal or diamonds, yet every sample of coal and diamonds had carbon-14 present! The Earth is thousands of years old, not millions. The Earth really is the recent creation of God.

**I will remember the works of the LORD: surely I will remember thy wonders of old. I will meditate also of all thy work, and talk of thy doings.**

**~ Psalm 77:11-12**

Scientists recently discovered that many dinosaurs ate grass. They came to this conclusion by analyzing coprolite (fossilized feces). This should be a problem for evolutionists because for decades we have been taught in textbooks and evolution charts that grasses had not evolved until 55 million years ago. This would have been millions of years ***after*** the extinction of dinosaurs (around 65 million years ago).

So how did that grass get into the petrified dinosaur dung? No problem for evolution – they just re-adjust their belief system to "evolve" grass earlier. Since evolutionism is a belief rather than science, it can be molded and stretched to fit any new observation which comes along. Contrast this with the solid foundation of truth provided by the Bible – which predicts in advance these kinds of observations. The Bible states that all plants (even grass) were made on day 3 of the creation week, while dinosaurs were made on day 6. There have been lots of "within-a-kind-variation" but no evolution of totally new kinds of plants or creatures, i.e. grass was available for dinosaurs to eat from the very beginning of creation. God made everything fully formed during the creation week – a fact which scientific observation continues to affirm.

**The grass withereth, the flower fadeth: but the word of our God shall stand for ever.**

**~ Isaiah 40:8**

Scientific research has revealed that dogs have at least six distinctly different barks that communicate their feelings to humans. Ranging from playful to sad, threatened to contented, angry to pleading, the frequency and pitch of a dog's bark is capable of communicating its feelings to its owner. In addition, most dog owners acknowledge that this communication goes both ways. Dogs seem to have the ability to sense and empathize with the mood of their owners, providing comfort to humans when they are sad, depressed, or lonely. One Border Collie has been shown to



understand and identify over 300 different objects based purely on spoken communication.

Shortly after the creation of Adam, God paraded all of the different creatures He had made before Adam and allowed Adam to participate in creation by naming the animals. It was after this that God brought woman, created from Adam's own body, to become Adam's companion – completing and sharing the glory of God's creation with him. No animal can replace the fellowship mankind desires from other people, but dogs do seem to have a special bond with humans.

It seems possible that the special bond between humans and dogs is representative of the original bond God intended between humans and other creatures. Before sin entered creation, mankind may have shared this type of close relationship with all animals. The bond between humans and dogs provides a hint of the bond we were created to have with all forms of life upon our planet.

**And Adam gave names to all cattle, and to the fowl of the air, and to every beast of the field; but for Adam there was not found an help meet for him.**

**~ Genesis 2:20**

DNA is a language:

- A language is a series of non-repeating symbols that carry a meaning. DNA is a series of non-repeating chemicals that carry information.

- A language causes a response by those receiving the message. DNA causes responses to happen throughout the cells within our bodies.

- Language has structure provided by spaces, punctuation, and divisions in the text. DNA is structured in this way.

- A language transfers information. DNA is the most compact information transporting system in the universe.

- Most importantly, a language has an author. No information can exist without an intelligent source.

Who authored DNA? Who put the unbelievably complex design and unfathomable quantity of information upon the DNA code of



every human being? The Bible gives us the answer. One of the many names of Jesus Christ is the "author of life". He is the ultimate author for he wrote the very language found upon the DNA molecule – the language of life.

**And killed the Prince of life, whom God hath raised from the dead...**

**~Acts 3:15**

What is warm-blooded, sits on its eggs like a hen, and has six legs? A bumblebee.

Have you ever noticed a bumblebee flying about on a frosty morning? It can do this because it is warm-blooded – a rarity among insects. Even if the air temperature is near freezing, a bumblebee can maintain a body temperature of 95°F. The bumblebee's blood is heated in the middle section, or thorax, of its body. When the temperatures are cool, it shivers its flight muscles, creating heat. Then the heat is circulated through her narrow waist



into her abdomen, warming her up. Yes, a bumblebee has a small waist just like a wasp but it is covered with hair. On a hot day, bumblebees can overheat. To prevent overheating, large air sacs in the abdomen inflate and deflate, allowing air to escape through spiracles (or tiny holes).

When the queen bumblebee wants to lay her eggs, she first gathers flower nectar. Then from her abdomen, thin sheets of wax are extruded. She collects these with her feet and brings them to her mandibles to chew, shape and form into a cell or 'pot' in which to put honey, which she will eat while incubating her eggs. Eating this honey will reduce the amount of time she needs to spend searching for nectar away from the eggs. Once she has this food store, she gathers pollen and forms a ball of pollen mixed with honey. On this ball she lays 4-16 eggs. She extrudes more wax and covers the pollen and eggs. Next she sits on her eggs like a setting hen. She lies with her abdomen gently over the wax covered ball, keeping it at about 86°F. The warmth of her body causes the eggs to hatch in 6 days, and after 21 days young bumblebees emerge.

There is no such thing as "simple life". God in His wisdom created all these fascinating creatures!

**I will speak of the glorious honour of thy majesty, and of thy wondrous works.**

**~Psalm 145:5**

MARCH

*God honors him who honors God*

**- Eric Liddell** (1902-1945)
Scottish 400M Olympic gold medalist in 1924,
martyred missionary to China

*But seek ye first the kingdom of God,*
*and his righteousness;*
*and all these things shall be added unto you*

**- Matthew 6:33**

Evolutionists believe that birds evolved from reptiles. But birds have many distinct features that are completely different in structure from land-dwelling animals. Consider just the shape of a bird's wing. Wings can only function if they are fully developed. In other words, a bird with half-formed wings cannot fly. How could the front arms of a reptile change into perfectly functioning wings? And it is not just the wings that would have had to evolve. For an animal to fly it would need lighter bones, a different lung system, different muscular and skeletal systems and a very specialized heart circulatory system. Everything about the reptile would need to be changed before flight could take place…and all these mechanisms would have to exist at the same time.

Before evolution should be taught as a fact, a fossil record showing the transition of half-winged/half-legged or half scaled/half-feathered creatures should be presented. The closest evolutionists have come to explaining the "evolution" of birds is a fossil called archaeopteryx. They believe this animal was a reptile in the process of



turning into a bird because this fossil has socketed teeth and claws on the wings. Yet other modern birds exhibit these same features, and archaeopteryx had fully formed wings, feathers, and a bird-like sternum for wing-muscle attachment. Archaeopteryx was fully a bird or a fraud! God created birds fully formed on day 5 of creation week, while reptiles were created **after** the birds on day 6.

**I will teach you by the hand of God: that which is with the Almighty will I not conceal.**
**~ Job 27:11**

Niagara Falls is the most powerful waterfall in North America. Every minute, four million cubic feet of water flows over the Falls. Niagara Falls has created a 7 mile (37,000 feet) long gorge via water erosion. A number of years ago, engineers took steps to slow the Falls from eroding westward. Prior to that time, the Falls was traveling backward about 4-5 feet a year. This rate and simple division indicates that the Falls is between 7,000 to 9,000 years old (37,000 feet divided by 4 or 5 feet per year). But is this the correct date?

Several factors could affect the erosion. What if there was more water in the past? What if the rock layers were more easily eroded shortly after the rock layers formed? Centuries after the Flood, there could have been much more water and the recently deposited rock layers might have been softer and more easily eroded. We cannot actually prove the age of the Falls. Evolution



promoter Charles Lyell visited Niagara Falls in 1841 anxious to find geological features which disproved the biblical time frame for creation and the Flood. His goal was to cast doubt upon the Bible, and he stated that the Falls eroded only one foot per year (even though he was told otherwise), thus dating the falls to be 37,000 years old. This caused many to abandon the Scripture's chronology and accept evolutionary concepts for geology.

When we understand the actual erosion rate for Niagara Falls (4-5 feet per year), we find that the Falls, sediment/rock layers, and gorge had to have been recently created. Contrast this with the age required by evolution. According to evolutionists, the sediment and rock through which Niagara Falls has eroded the gorge are millions of years old. So why is the gorge only thousands of years in the making? Obviously the Biblical time frame fits the data much better. The Bible stands true – as always!

***Who is this that darkeneth counsel by words without knowledge?***

**~ Job 38:2**

During the development of a baby inside the mother's womb, it is covered with a fine coat of downy hair called lanugo. This hair develops for a brief period when the pre-born baby is six months old and then disappears. Since the mother's womb is a cozy 98.6 °F, this hair is obviously not needed to keep the developing baby warm. Thus it has been cited as a useless feature and labeled as a "remnant of our primate ancestry".[1]



In reality, this fine downy hair is absolutely critical for the development of the baby within the mother's womb. Just after the lanugo grows on the baby's developing skin, the baby produces a waxy coating over its skin called the vernix caseosa (meaning "cheesy varnish"). It is this waxy coating that protects the developing skin from the fluid in the mother's womb. Without the fine coating of hair, the waxy varnish would have nothing to anchor itself to and could not function to protect the underlying skin.[2]

So, once again, the evolution theory has proven to be both wrong and an impediment to knowledge. If widely accepted that the lanugo were a useless leftover feature from the past, there would be no reason to search for its function. Furthermore, the fact that recently published books on evolution still stoop to presenting this feature as evidence for evolution[1] (in spite of widely acknowledged medical knowledge to the contrary) is an indication of how little evidence for human evolution really exists.

**... Thou hast covered me in my mother's womb.**

**- Psalm 139:13**

Bird migration is more complex than planning a family vacation. Birds start preparing themselves for fall migration as the amount of daylight shortens. They go into a feeding frenzy in order to gain as much weight as possible. Then they begin to gather into groups, experiencing migratory restlessness. At just the right moment, all over the world, billions of birds take off on perilous migratory flights. How do they know their destinations? Scientists have shown that birds raised in captivity, completely separated from parents, still know when and where to migrate. It is actually programmed **into their DNA!**



Scientists are discovering multiple migration mechanisms:

- Birds use landmarks such as rivers/mountains/seashores.
- Some birds use the "sun compass", which requires an internal clock. The sun's movement and the bird's internal clock allow them to calculate their direction.
- Many birds migrate primarily at night when there is less wind and fewer predators. These birds use a "star compass". If the birds migrate beyond the equator into another hemisphere, they need to know the stars in that hemisphere as well.
- What if it is cloudy? Birds seem to have the ability to sense and navigate by the Earth's magnetic field. Birds have been found to have small amounts of magnetic crystals in their heads, i.e. they have a built-in compass!
- Birds have been found to use polarized light, ultraviolet light, air pressure differences, odors, and low frequency sounds made by the wind and sea to aid their migration.

Since more than one navigational method is required for successful migration, the problem for any evolutionary explanation is multiplied exponentially. Even one navigational method programmed into a bird's DNA is amazing but God has provided multiple backup methods. This is testimony to a Creator whose awesome power and intellect reveals that no detail is too small for His consideration.

**"I am the way, the truth, and the life:**
**no man cometh unto the Father, but by me."**
**~ John 14:6**

Have you considered what the Christian worldview has produced:

- Christians started most hospitals, hospices, and medical clinics.
- Florence Nightingale received inspiration from Jesus Christ.
- Nearly all of the first 123 American universities were started by



Christians to teach others how to learn from…THE BIBLE!
- Before the civil war, 2/3 of abolition societies were headed by pastors.
- Before Christianity, cannibalism was widespread, and Anglo-Saxons drank human blood – it was the Gospel which civilized barbaric cultures.
- Modern science – The vast majority of scientific discoveries are from Christian-based countries.
- Capitalism and free-enterprise – Christianity is directly responsible for work discipline, self-reliance, and self-denial.
- Elevation of women. – No other religion elevates women to equality with men.
- Representative government – Patrick Henry stated, *"It cannot be emphasized too strongly or too often that this great nation was founded, not by religionists, but by Christians; not on religions, but on the gospel of Jesus Christ!"*
- Civil liberties – *"All people are CREATED equal."* (American Declaration of Independence)
- Benevolence and charity – YMCA-Young Men's Christian Association, Salvation Army, Red Cross, A.A., Teen Challenge.
- Great works of art and music (produced by those wishing to bring Glory to God).

The Christian worldview has made more changes on Earth for the good than any other movement or force of history. These are just a few of the positive contributions that Christians have made throughout the centuries. God is concerned for our life not only in eternity, but also on Earth, and the Christian worldview, when practiced, brings forth goodness.

**But the fruit of the Spirit is love, joy, peace, longsuffering, gentleness, goodness, faith, meekness, temperance: against such there is no law.**

**~ Galatians 5:22,23**

After the Civil War, the Southern soils were worn out from planting only cotton, which depleted the soil of many minerals



and necessary nutrients. Farming was terrible, and farmers could barely make a living. That was the sad situation until one American scientist turned things around for an entire region of our country.

George Washington Carver made it his life's work to help farmers, and he achieved success beyond his wildest dreams. Dr. Carver searched for a plant that would revitalize farming in the South. He settled on the lowly peanut; this plant grew wild in the South, and it was considered worthless pig food. At this time the peanut was a nuisance weed, little better than a yard full of dandelions…until Carver began working with this "nuisance weed".

He patiently analyzed the peanut and broke it down into compounds. He then began to put these compounds back together in various combinations until he had produced more than 300 different products ranging from soap to shampoo, meat sauces to milk substitutes, and, of course, peanut butter and peanut oil. The peanut was also a legume that restored nitrogen to the worn-out soil. The lowly peanut, which had been only good for pigs, became a billion-dollar industry.

What is most interesting is how Dr. Carver began each day. He rose at 4 a.m. and took a walk. During these morning walks, he prayed for guidance and help for the day's work. When success came, he gave God the glory. He once said, "Without God to draw aside the curtain, I would be helpless…" He called his laboratory "God's little workshop." He believed that God made the world, and it was God's delight to reveal the secrets of creation to those earnestly searching for them.

**It is the glory of God to conceal a thing:**
**but the honour of kings is to search out a matter.**
**~ Proverbs 25:2**

The sheer number of varieties of plants staggers the imagination. It is estimated that there are more than 100,000 plant species on Earth; grasses alone have more than 5,000 different forms. Even within a given species, the variety is never ending; there are over a million different varieties of corn or grapes.

When God spoke plants into existence on Day 3 of creation week, there was a cacophony of variety created – grasses, herbs, and fruit immediately covered the Earth (1:11-13). The first plant mentioned (grass) has a seed that is not particularly obvious to the eye until maturity. The second plant mentioned (the herb) has seeds apparent as a visible feature. The third plant has the seed in the fruit. Thus the writer cataloged the plants by a simple, natural division- using the structure of the seed/plant as the guide. The Bible provides the first broad classification of plant groupings.

God also made sure that we understand that organisms reproduce "after its kind". The expression "after its kind" occurs ten times in the first chapter of Genesis. God decreed a boundary between very different kinds of creatures. There can be change within a kind, but to change one type of plant or animal into a completely different kind is simply impossible. The study of genetics has shown this to be absolutely true.

The biological world is like a musician who combines the notes in countless numbers of ways and produces a great number of different harmonies. But the musician can only make harmonies using the notes that are present. The number of variations is restricted by the number of notes available. In like manner, the variety produced in a given kind is limited by the number of possible combinations of genes available.

**And God said, Let the earth bring forth grass, the herb yielding seed, and the fruit tree yielding fruit after his kind....And the evening and morning were the third day." ~ Genesis 1:11, 13**

There are over 100 "clocks" that we can examine to get an idea of how long the Earth has existed. Almost all of these "time pieces" show an Earth that was created recently.

Here are but a few:

- Earth's magnetic field is decreasing. Every 1,400 years, the field strength decreases by half. If you go back 6,000 years, life is fine, but if you go back more than 10,000 years, Earth would have had the field of a magnetic star. Life on Earth would not have been possible.
- Our oceans are becoming saltier each year. If the oceans of the world were even a billion years old, they would be as dead as the Dead Sea.
- The Mississippi River dumps tons of sediment into the Gulf of Mexico each year, creating a river delta. The size of the delta can be explained in a mere 5,000 years. It should be much



larger if the Mississippi River is millions of years old.
- Most places on the planet average about 6 inches of top soil. This much soil from decayed vegetation can be accounted for in about 5,000 to 20,000 years. We should have WAY more top soil if Earth is billions of years old.
- What is the oldest living thing on Earth? The bristle cone pine tree is approximately 5,000 years old. What happened about 5,000 years ago? The Flood of Noah – after which this seed was planted and has been growing ever since.
- Continents are eroding. All continents should have been eroded to sea level in less than 100 million years. Yet, we still have continents. The Earth cannot be billions of years old.

Creation scientists have documented well over 100 "time clocks" that support a recently created Earth. The Earth cries out, "I am young."

**The north and the south thou hast created them:**

**~Psalm 89:12**

There are approximately 200,000 different proteins that make up the majority of the structures within the human body. As far as we know, every one of these proteins has some useful, functional purpose. Why are our bodies not full of useless leftover chemicals from our evolutionary past? These proteins vary from the crystal clear protein solution within our eyeballs to the enamel on our teeth. Starting out as a single cell (dividing and doubling in size approximately every 30 minutes) and forming increasingly complex structures, our bodies know how, where, when and how much of each of these proteins to form. Too much and we die. Too little and we die. Tissue produced in the wrong location and we die.

Most amazing is how each protein is produced. They are formed from coded information upon the DNA molecule. But to form each protein, a little protein machine must attach to the DNA at just the right point to start the process. **Then a different protein machine must start the unzipping process.** Next, a third protein machine must line up a special molecule (RNA) of exactly the right length, and a fourth protein must know when to stop the copying process. The RNA passes through a fifth protein machine that allows an exact copy of the information to be made.

Here is the trillion dollar conundrum. Biologically useful proteins **NEVER** exist in nature unless DNA makes them, and DNA cannot be made, or unzipped, without these proteins. So, where did either one come from? Just as you can't have a chicken



without an egg, you can't have an egg without the chicken (chickens were formed with eggs already inside the female chicken). The obvious answer is that the entire system was created simultaneously – yet this is the one answer not allowed to be considered by students.

***Before I formed thee in the belly I knew thee; and before thou camest forth out of the womb I sanctified thee... ~ Jeremiah 1:5***

The idea that the universe had a beginning is a profoundly Christian concept with significant implications. A "beginning" of the universe implies that there was no universe before "the beginning", and time does not extend indefinitely into the past. It also implies that things are running downhill, and that time will not extend indefinitely into the future. A beginning implies an end.

The very existence of the universe is an enormous problem for modern physics. The first and foremost law of physics is that "matter and energy can be neither created nor destroyed". Yet the universe is full of both. Atheists are therefore forced to believe that "nothing" somehow turned into "everything" – which violates the most basic law of science. The second law of thermodynamics states that every process results in a loss of useable energy. Hot always flows to cold…natural processes tend toward increasing randomness. Thus, if the universe were infinitely old, there could be no useable energy left, and everything would be a uniform temperature and evenly dispersed. The universe cannot possibly have been in existence for eternity. Since the universe cannot be eternal, and it cannot have made itself, the only logical conclusion



is that something outside of the universe made the universe, i.e. God.

In summary, the universe had a recent beginning (it cannot be infinitely old) and a Creator must exist. Furthermore, it is this Creator, not the universe, which is eternal, and there will be an end – a wrapping up of time – in this current physical universe.

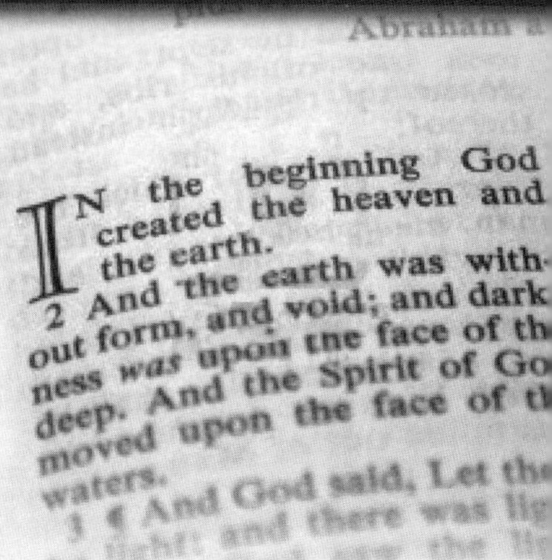
IN the beginning God created the heaven and the earth.
2 And the earth was without form, and void; and darkness was upon the face of the deep. And the Spirit of God moved upon the face of the waters.

**For this God is our God for ever and ever: he will be our guide even unto death.**

**~Psalm 48:14**

### Evolutionary view of Self-Esteem

You are the descendant of a tiny cell of primordial protoplasm that washed upon an ocean beach 3 ½ billion years ago. You are the product of time, chance and natural forces. Your closest living relatives swing from trees and eat crackers at the zoo. You exist on a tiny planet in a minuscule solar system in an obscure galaxy in a remote corner of a vast, cold and meaningless universe. You are flying through space with no purpose, no direction, no control and no destiny. Your final outcome: destruction. All you are is a bunch of chemicals and once you die, you will cease to exist. In summary: you came from nothing, you are going nowhere, and your body will end up beneath 6 feet of dirt where it will become food for bacteria and worms. Now, don't you feel special?

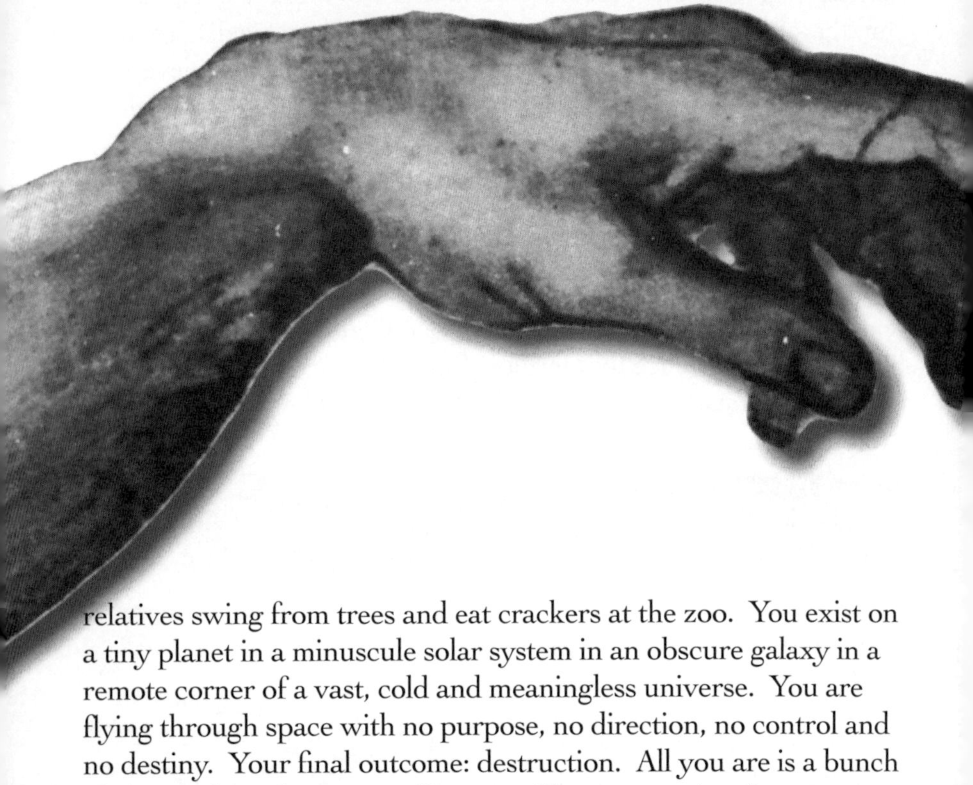

*There is a way which seemeth right unto a man, but the end thereof are the ways of death.*
*~ Proverbs 14:12*

## Christian view of Self-Esteem

You are the extraordinary creation of a good and all-powerful God. You are the climax of His creation – created in His image – with the capacity for creativity, feelings and worship. You are not just an animal. You are individually distinctive among all other people. God has masterminded the exact combination of DNA and chromosomes that constitute your genetic code, making you as unique as one

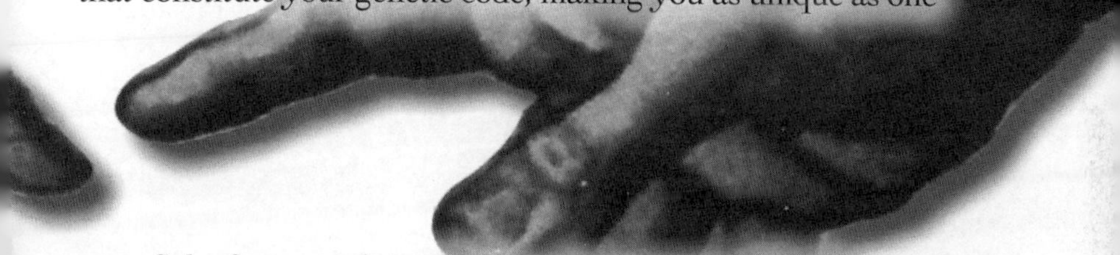

snowflake from another. God has given you special gifts and abilities to serve Him in a unique way. In spite of indifference, rebellion, or disobedience, your Creator intensely desires your companionship. He gave the life of His one and only Son, Jesus, so that you might spend eternity with Him. However, you must accept that gift. As a Christian, you are clothed with the righteousness of Christ and given the privilege of eternal life in God's presence.

Your heavenly Father is sovereign and will allow nothing to cross your path that is not Father-filtered. He cares for you so much that He is totally available to you at all times, and He listens to every word you say. He cares deeply about your hurts and has a plan for your life. He has given you the inspired Word of God as a road map for living. Your destiny is to live forever in a magnificent kingdom and enjoy the wonders of both His presence and His creation. Now, do you feel special?

The question of origins is not just a dry academic subject; it determines your very perspective on life and how you choose to live it.

**"For I know the plans I have for you," declares the LORD, "plans to prosper you and not harm you, plans to give you hope and a future."**

**~ Jeremiah 29:11**

Scripture teaches that one of the purposes of the night sky is to reveal God's glory. This is especially apparent when observing Pleiades and Orion. Orion, the hunter, is probably the best known constellation after the Big Dipper. This constellation is easy to trace in the night sky; we can see the shape of a man. Orion is easily found by the three bright stars forming his belt. One of Orion's arms is raised high holding a club while the other holds a shield. Facing Orion is Taurus the Bull, which includes the beautiful star cluster Pleiades or the Seven Sisters.

These two star patterns, Pleiades and Orion, are mentioned in the Bible when God asked Job, "Can you bind the chains of Pleiades,

or loose the cords of Orion?" (Job 38:31) One star system is said to be bound together, and the other is said to be separating. It is only recently that astronomers have realized that the stars in the cluster known as Pleiades are gravitationally bound to each other…while the stars in the belt of Orion are flying apart. This was written about 4,000 years ago, and only recently have we found this to be accurate. Because the Bible is not a book of opinion but is the inspired revelation of the Creator of the universe, whatever the Bible states concerning the physical world (science) always turns out to be true.

**Canst thou bind the sweet influences of Pleiades,**
**or loose the bands of Orion?**
**~ Job 38:31**

How do bees know if there is nectar in a flower? Many flowers provide a visible sign on their petals. Consider the small desert lupine flower; it would take a bee hundreds of visits to get enough nectar to make honey from this small flower. As the bee flies along, it notices the bright yellow spot on the desert lupine's blue petals. This bright yellow spot says, "Here's nectar!" So the bee will stop in and get the nectar. But after it leaves, the yellow spot turns black. When another bee comes along, it sees the black spot and knows the nectar is gone!

## INSPIRED BY BIOLOGY

MARCH 14TH

Somehow bees know to stop only at the flowers with yellow spots. This saves the bee lots of time and energy. When a bee picks up its nectar, it also pollinates the plant. If the flower has already been pollinated, there is no need for it to be pollinated again. Other flowers provide similar clues. When the nectar is gone from the flowers of the vetch, the petals show black splotches.

Many flowers advertise on their petals using splotches and spots saying, "Here's nectar", and when there is no nectar, "Don't stop here, the nectar is gone." How did this come about? Both the bees and flowers would have to cooperate to get this to happen. This is design, not random chance.

**To the only wise God our Saviour, (be) glory and majesty, dominion and power, both now and for ever.**
**~ Jude 1:25**

It does not take millions of years to make petrified wood – just a short period of time and the right conditions. Biblical geologists have claimed this for years, and it has been recently confirmed by five Japanese scientists with their observations published in a secular geology journal. These scientists, led by Hisatada Akahane, studied a small lake cradled in the crater of the Tateyama Volcano in central Japan. The crater is filled with steaming acidic waters, which gush from the bottom of the lake. This mineral-rich solution fills a 35 foot pond with a waterfall that cascades over the edge. Fallen wood trapped in the overflow was found to be heavy and hard; it was totally petrified with the mineral silica. Surprising to old-earth geologists, this petrified wood was only 36 years old.

These scientists then experimented by fastening pieces of fresh wood on wires and lowering them into the lake. After seven years, they were found to be petrified – the carbon of

the wood had been completely replaced by silica and the wood had turned to stone. All it takes to turn wood to stone is hot, mineral-rich waters.

It does not take millions of years. The flood of Noah's time would have offered the right conditions, plenty of mineral-rich water and volcanic activity. No wonder petrified wood is relatively common in rock layers around the globe.

***Deep calleth unto deep at the noise of thy waterspouts: all thy waves and thy billows are gone over me.***

***~ Psalm 42:7***

The Grand Canyon is located in the wrong place. As the Colorado River flows across the flat plains, upstream of the Grand Canyon, it encounters a vertical plateau of rock rising 3,000 feet above the river. Instead of flowing around this wall of rock and continuing south, the river makes a ninety degree turn to the west and carves right through 3,000 vertical feet of solid rock at the highest point of the rock "dam". This is an enormous mystery to those who believe the river carved the canyon over millions of years – after the Kaibab Plateau was upraised. Here is a typical acknowledgement of this astounding problem: *"Oddly enough, the Grand Canyon is located in a place where it seemingly shouldn't be. Some twenty miles east of Grand Canyon Village, the Colorado River turns sharply ninety degrees, from a southerly course to a western one and into the heart of the uplifted Kaibab Plateau…It appears to cut right through this uplifted wall of rock, which lies three thousand feet* **above** *the adjacent Marble Platform to the east."* [1]

Either enormous amounts of water broke through the recently laid sediment and catastrophically carved the canyon as backed-up water covering a large area of the western United States drained through the area…or the erosion rate of the river EXACTLY matched the slow, gradual uplift of the 3,000 foot plateau over millions of years.

It would seem to take far more faith to believe millions of years of perfectly matching uplift/erosion than to simply accept the much more straightforward evidence for rapid erosion following the flood of Noah 4,500 years ago. The biblical model explains things better; it is primarily ignored, however, because it eliminates the time needed for evolution to have happened.

**In thee, O Lord, do I put my trust…for thou art my rock and my fortress.**

**~Psalm 71:1, 3**

Have you considered the shamrock of St. Patrick's Day? St. Patrick is the patron saint of Ireland who was born 389 A.D. in Roman Britain. At about the age of 16, Patrick was captured by pirates, sold as a slave in Ireland, and languished in hard labor for the next six years. During this time, he remembered his father's words about the one true God. In the midst of his solitude, he sought this God for his life. While serving as a slave, Patrick confessed his waywardness and sins, and although he had nothing to offer, he devoted his life to serving Jesus. Later, he had a dream, and in this dream a voice called out, "Lo, thy ship is ready!"

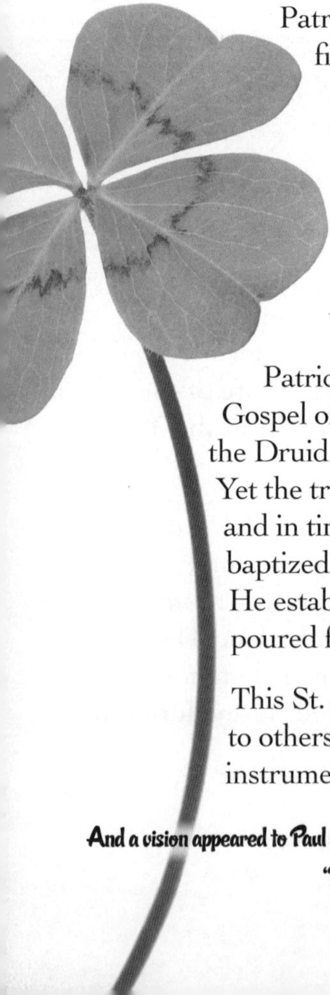

Patrick fled 200 miles, arriving at the sea to find a ship willing to give him passage off the island. He sailed to France and after a few years returned to Britain. But he could not take his mind off the Irish Druids (nature worshipers). Twenty years passed, and Patrick had another vision, this one of the Irish calling out to him, "Come here and be with us".

Patrick returned to the island proclaiming the Gospel of Jesus Christ. But this was no easy task, for the Druid priests tried to kill him on multiple occasions. Yet the truth of God's love could not be suppressed, and in time a great revival began. Patrick personally baptized 120,000 of the 300,000 inhabitants of Ireland. He established 300 churches from which missionaries poured forth onto the European continent.

This St. Patrick's Day proclaim the one true God to others. Perhaps Jesus wants to use you as the instrument for transforming an entire culture.

**And a vision appeared to Paul in the night; There stood a man of Macedonia, and prayed him, saying, "Come over into Macedonia, and help us."**

**~Acts 16:9**

The koala feeds on eucalyptus leaves - which happen to be poisonous. Yet the koalas make their homes in these very trees, eating the poisonous leaves for nourishment and water. As a matter of fact, that is all koalas will eat. When Koalas are put into zoos, great quantities of the poisonous leaves must be shipped in to keep them alive. The koalas eat so many eucalyptus leaves that they take on the smell of a eucalyptus-flavored cough drop.

How does the koala survive while eating poison all day long? The answer is its digestive system. Koalas, like many other mammals,

cannot digest the cellulose in the leaves; therefore, they depend on micro-organisms within their intestines to digest them. The place where the small and large intestine converge is the caecum. The caecum functions as a fermentation chamber where microbes digest cellulose. Unique to the koala bear is a specially designed caecum containing bacteria that neutralize the poison found in eucalyptus leaves.

Why would these poison-neutralizing microbes exist in the koalas if they were not eating these poisonous leaves? And even if the koala had these microbes, which koala would volunteer first to eat these leaves when other creatures died every time they ingested the leaves? This caecum, with the neutralizing microbes, and the koala bear's instinct to eat poisonous leaves had to have been created at the same time. It could not have happened by chance. God designed it. Isn't it wonderful to see God's fingerprint in the "cough drop" bear?

**And by the river upon the bank shall grow all trees...whose leaf shall not fade...and the fruit thereof shall be for meat, and the leaf thereof for medicine. - Ezekiel 47:12**

Raindrops are made when water molecules begin to stick to tiny condensation nuclei such as smoke, pollen, salt grains, or dust within clouds. The droplets grow as they collide with more and more water molecules. Finally, they are heavy enough to leave the clouds and fall as rain. Job wrote of this exact process almost 4000 years ago! Raindrops do not look like teardrops but are spherical in shape. As the drops become larger and larger, more and more air resistance builds up, causing the spherical shape to flatten out a bit, which makes them look more like tiny hamburger buns falling from the clouds.

Raindrops are limited in their size. They have a tendency to disintegrate when they exceed about ¼ inch in size. The raindrop's surface tension prevents the drop from growing too large as they fall to the Earth. Imagine if raindrops could grow larger and larger; they might be as large as bowling balls or pianos. Look out below!

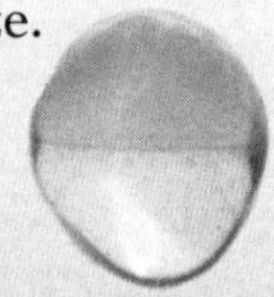

Imagine the erosion which would take place if the ground were pummeled with piano sized rain drops! Imagine those large raindrops hitting the plants and trees; they would certainly be torn and shredded! On the other hand, what if the raindrops were too small, like fog droplets – rain would not reach the ground but remain suspended in air currents. Raindrops too large or too small wouldnot be good for the Earth. Who would have thought that even the size of a raindrop falling

to the ground was important? God did. So the next time it rains, thank God for just the perfect sized raindrops.

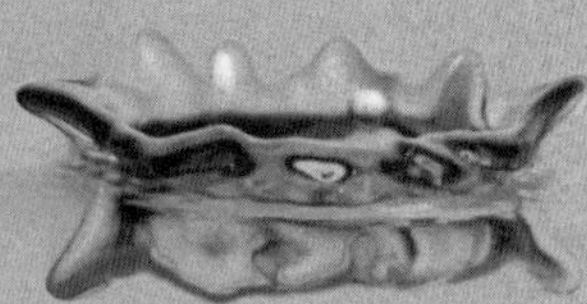

***For he maketh small the drops of water: they pour down rain according to the vapour thereof: Which the clouds do drop and distil upon man abundantly.***

***~ Job 36: 27-28***

Most of the time, pain is a good thing. It warns of sickness or injury. If you feel intense pain, you rush off to the doctor. If you never felt pain, you would rarely slow down. Pain warns us. However, in life-threatening situations, pain can hinder survival.

There are true life survival stories where people have been severely injured yet not incapacitated by pain. Near the edge of a cliff, a



hiker finds the ground crumbling underneath and falls 60 feet onto a granite slab. Awaking from the fall, the hiker is badly hurt but not overwhelmed with pain. A soldier is badly wounded in battle yet keeps on fighting – feeling no pain. In each case, survival is needed, so the brain temporarily blocks the pain. How can our brains do this?

It seems God has given our brain a contingency plan, ready to activate at a moment's notice when we are faced with a life-threatening situation. Our brain recognizes the threat and initiates steps for our survival. Scientists do not fully understand the details, but they call this "the gate-control mechanism." Nerves at the injury site send signals along the nerves to a projection neuron (the gate) located on the spine that then forwards the message on to the brain. If the pain must be blocked, the periaqueductal gray in the brain closes the gate by releasing natural pain killers. These natural pain killers are more powerful than morphine. Once the danger has passed, the periaqueductal gray removes the natural pain killers, allowing pain through the gate.

We never know what unexpected dangers might affect our lives. But God knows, and He has equipped us with back-up systems, so we have the best chance to survive.

***And God shall wipe away all tears from their eyes; and there shall be no more death, neither sorrow, nor crying, neither shall there be any more pain...***

***~ Revelation 21:4***

How do butterflies survive freezing winters? We often see butterflies early in the spring that could not have come from caterpillars. Why don't the butterflies freeze and die?

The Mourning Cloak butterfly hibernates in holes or behind loose bark. But before it goes to sleep for the winter, it fills its body with antifreeze. Normal cells are filled primarily with water, but water expands 9% when frozen. This expansion within a cell would cause the cell membranes to break – killing the creature. So the butterfly synthesizes glycerol within its cells, allowing the butter-



fly's body to cool to -50°F without freezing. In early spring this butterfly re-appears, even before most other insects. The butterfly feeds on the sap of spring trees, which leaks from injuries and buds.

The Mourning Cloak butterfly also has mostly black on its wings that act like solar collectors absorbing the heat from the sun; this allows this cold-blooded insect to move around readily in the still cool springtime. This butterfly is not the only one that hibernates – there are several other butterflies and many other creatures that make antifreeze in the fall and hibernate through the winter.

How did the first Mourning Cloak butterfly learn to make an anti-freeze? God designed this butterfly to survive the brutal winter as an adult butterfly. The next time you see a butterfly very early in the spring – chances are it is a butterfly that was designed to make antifreeze!

***...there is no other God who can deliver like this.***

***~ Daniel 3:29***

Man has been intelligent from the beginning of time. Here are just a couple of examples:

## INSPIRED BY HISTORY

March 22nd

- In 1900, near the island of Antikythera, Greek divers found an ancient shipwreck from about 65 B.C. Recovered from the wreck was a complex mechanical computer. The "Antikythera Mechanism" used up to 72 bronze gears to compute the motions of the moon, sun and planets. It also used a differential gear concept which was not "invented" until the 17th century A.D.
- A strange relic made by the Parthians (250 BC - AD 224) was recently found in Southern Iran. Inside a simple earthenware jar was a copper cylinder with an iron rod inside. The parts were soldered with a 60/40 tin/lead alloy and were cemented with asphaltum. When a General Electric engineer made a replica and added grape juice, it immediately began producing a steady half-volt of electricity. Using this current, he electroplated a silver statuette with gold. It was not until 1799 that Alessandro Volta "discovered" how to make an electrical battery and the early 1800's that Michael Faraday "invented" electroplating.

We regularly read reports of amazing accomplishments of ancient civilizations because ancient peoples were not primitive brutes but extremely brilliant and innovative people. Adam and Eve were created highly intelligent by God. Adam named the animals before Eve was even created. That means he observed the characteristics of each animal and instantly gave each a suitable name. God made man intelligent from the beginning.

**From the place of his habitation he looketh upon all the inhabitants of the earth. He fashioneth their hearts alike; he considereth all their works. ~ Psalm 33:13-15**

Have you truly appreciated your dandelions – those unstoppable weeds which plague your perfect lawn? How does this weed send out seeds and spread so rapidly? Each plant sends out hundreds of seeds - carried away by the wind to other parts of your lawn – like miniature parachuting commandos. A strong wind can carry these seeds for miles. But how do they get their miniature parachutes airborne? To be effective, the seeds need to be released high above the grass, yet dandelion leaves hug close to the ground so that photosynthesis, plant food production, can continue. Or maybe they just know they need to stay low so that the mower can't chop them off!

Have you noticed that when it wants to seed, the dandelion sends up a tubular stem that towers above the grass? Even if you just mowed it down the day before; up, up and away goes the stem and the dandelion parachutes – flying all over your lawn. The master Designer knows how to produce a design that works!

One other thought on these supposedly "useless" weeds. If grown in pesticide-free lawns, the young, tender leaves of dandelions are an easily-absorbed source of boron and calcium, quite healthy and delicious on your fresh salad.

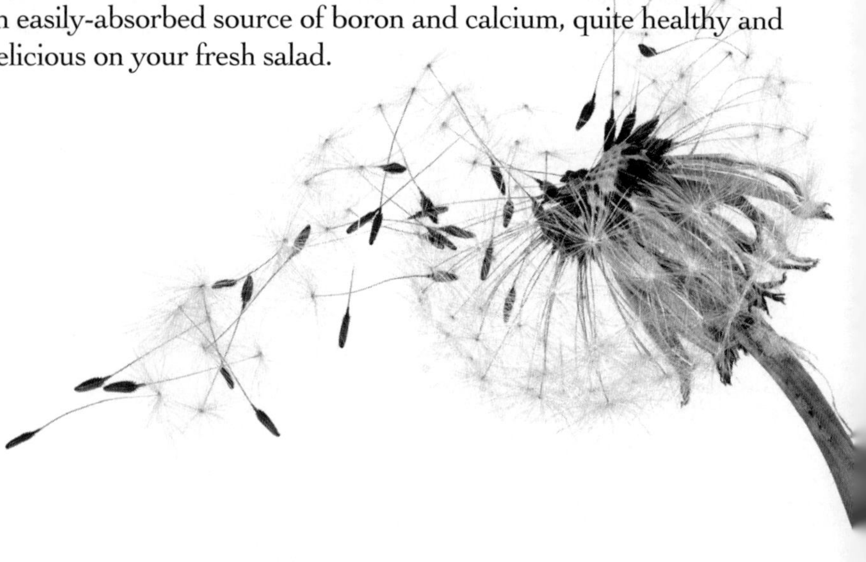

**Sing forth the honour of his name: make his praise glorious. Say unto God, "How terrible art thou in thy works!"**
**~ Psalm 66:2,3**

Have you considered that man is NOT an "animal"? College textbooks and nature centers often declare that "man is an animal." Evolutionary thinking reduces us to mere animals by promoting the idea that we are the product of random forces of nature. Yes,



we do have mammalian characteristics; however, we are not just animals. So what makes us different than animals?

- Man can be rational. He can construct arguments, use logic, and draw conclusions. Do you see lions in prolonged debates at lion universities?
- Man has remarkable languages with vast vocabularies, intricate grammar, and an ability to have deep conversations. Man even has the ability to write his language down. Do you ever see a duck writing books?
- Man has the longing to know why. He studies medicine, astronomy, math, etc. Do you see chickens pondering the universe?
- Man can recognize beauty. He can admire roses in a garden and a painting on the wall. Do you ever see a raccoon stopping to watch a sunset over a lake?
- Man is creative and imaginative. He develops music, dance, and drama. Have you ever seen cats performing Shakespeare?
- Man has a sense of humor. He delights in a good laugh. Do you see apes telling jokes?
- And what animal desires to worship its Creator?

Man is remarkably different than animals! Mankind was made in the image of God.

**For thus saith the Lord of hosts; After the glory hath he sent me unto the nations which spoiled you: for he that toucheth you toucheth the apple of his eye.**

**~ Zechariah 2:8**

Many theologians still accept an idea called "the gap theory"– popularized by Cyrus Ingerson Scofield (1843 - 1921) and promoted by notes in his Bible translation for almost a century. Scofield and subsequent theologians, influenced by the eighteenth century non-Christian geologists, "added" millions of years to the Biblical narrative. The logic of these theologians is as follows:

- If the Earth's rock layers are millions of years old, Noah's flood could not have created these sediment layers.
- Therefore, the record of death in these rocks must be the result of some previous tragedy.
- Therefore, there must be 'missing time' within the Bible's narrative of Earth history, i.e. a 'gap' in the Genesis narrative.

Scofield and others "found" this gap within Genesis 1:2 that they interpreted as a "re-forming" of the original creation from a state of chaos after the judgment of Satan via a 'pre-Adam' flood upon the Earth.

There are HUGE problems with this idea. First, it is TOTALLY non-credible to any geologist, so it does not reconcile the Bible to science. Second, it is TOTALLY speculative with essentially no direct textual support. Third, it attributes the rock record of the planet to a flood which is NOT mentioned in the Bible and TOTALLY ignores the flood which Genesis does talk about.



According to the gap theory, Noah's flood (which is described in the Bible) had no effect on the geology of the planet, while a mythical Lucifer's flood (which the Bible does not mention) did create all the geology of the planet! In an effort to make the Bible "more credible", Scofield and others ultimately destroyed its credibility.

**In the beginning God created the heaven and the earth. And the earth was without form, and void; and darkness was upon the face of the deep. And the Spirit of God moved upon the face of the waters.**

**~ Genesis 1:1~2**

In Harare, the capital city of Zimbabwe, there stands an 18 story shopping complex built with no mechanical heating or cooling system, yet the temperature inside remains comfortable year round. This design even uses 90% less energy as compared to a conventional building. How were the architects able to do this? The building is designed with a series of interior ventilation ducts that allow outside breezes to moderate the temperature throughout the building. The design of the building was inspired by the design of termite mounds found across the plains of southern Africa. They show God's design for building air movement.

## INSPIRED BY BIOLOGY

MARCH 26TH

These above-ground termite mounds can exceed 10 feet in height and hold colonies of 15,000 adult termites. Although the outside air temperature varies from 35-104°F, the interior temperature never fluctuates beyond 87 +/-1°F. This temperature is required to keep fungus, which is grown by the termites as their food source, alive.

How do the termites maintain such tight temperature control within their mounds? They constantly open and close an intricate series of ventilation vents located throughout the mound. The network of tunnels resembles the veins and arteries of our circulatory system and works to move air (and the corresponding heat) to maintain a perfectly tuned internal mound temperature.

Even brilliant architects can learn from these small creatures because the wisdom of their Creator is reflected by their abilities and instincts.

*Great is the Lord, and greatly to be praised; and His greatness is unsearchable.*
~ Psalm 145:3

We know much about dinosaurs because of the rapid burial and fossilization of their bones. The conditions had to be just right for this to have happened:

- So many dinosaur skin impressions have been found that scientists have excellent information on the surface appearance of some dinosaurs. In order for them to be preserved, oxygen could not be present, otherwise decay would take place. The dinosaurs had to be rapidly covered with enough sediment to stop decay as the skin impressions turned to stone.
- In Dinosaur Provincial Park, Alberta, Canada, scientists have estimated that there are more than 3,000 bone beds, some of which contain millions of bones. Bone beds are found worldwide and contain billions of bones. Scientists believe that bone beds are excellent evidence for very fast moving water that carried decaying animals along until they were buried and fossilized.



- Fossil dinosaur footprints are found worldwide. How long do your footprints stay on a beach before being washed away? Yet, we find thousands of dinosaur footprints in stone. This is evidence of rapid coverage by a worldwide flood and then fossilization.
- Dinosaur eggs are delicate yet we find fossilized eggs worldwide. To preserve eggs as fossils requires rapid burial under lots of sediment.
- Fossil dinosaur poop (coprolite) is relatively common, yet how long does poop last in nature? To fossilize, dinosaur poop would require rapid burial under lots of sediment.

Notice the common themes – lots of sediment, rapid burial, rapidly moving water, worldwide in extent? Taken all together, the worldwide findings of the fossilized dinosaur skin impressions, bone beds, footprints, eggs, nests and poop provide conclusive evidence for a terrible raging worldwide flood. This worldwide deluge is mentioned in the Bible, The Flood of Noah's time (Genesis 6-9).

**And the waters prevailed, and were increased greatly upon the earth; and the ark went upon the face of the waters.And the waters prevailed exceedingly upon the earth; and all the high hills, that were under the whole heaven, were covered.**

**~Genesis 7:18-19**

Our solar system is full of astronomical evidences indicating its recent creation:

- Reports of volcanic activity on the moon. If the moon is old, it should be cold.
- The moon is moving away from the Earth. If the moon is billions of years old, it should be much farther away. This would affect Earth's tides, which are vital for life. Tides cleanse the shorelines and keep the oceans' currents circulating. No tides, no life.
- Saturn's rings have been widening rapidly. Saturn's rings are young.
- If Neptune were billions of years old it should be "cold" and



lack strong wind movement. Yet in 1989 Voyager II found it have the fastest winds in the entire solar system. This observation is consistent with a young age, not billions of years.
- Neptune's rings have thick regions and thin regions. This unevenness indicates they cannot be billions of years old, since collisions of the ring objects would eventually make the rings relatively uniform.
- Jupiter's larger moons, Ganymede, Io, and Europa, have magnetic fields which should not exist if they were billions of years old.

These are just a few of the more than 100 different indicators that our universe is quite young. Why aren't these evidences mentioned and discussed in classrooms and museums? Because a young universe can only be explained with God, and this subject is not allowed within our educational system.

**He hath made the earth by his power, he hath established the world by his wisdom, and hath stretched out the heavens by his discretion.**

**~ Jeremiah 10:12**

The elephant's trunk is the ultimate "Swiss army knife". With its trunk, an elephant can breathe, rub an itchy eye, greet and hug a friend, pick up a 300 pound load, snatch a penny lying flat on the floor, locate a nearby friend or food, grab tasty leaves from 23 feet up in a tree, tear up food and place it into its mouth, knock down a tree, use it like a hose to suck up over four gallons of water per minute, take a shower by spraying water over its body, sense enemies, sprinkle dust and fling mud on itself as protection from biting flies or the hot sun, dig 10 feet deep for water, raise a warning, scratch his ear, wrestle with friends, swim underwater using it as a snorkel, and

make a wide variety of sounds by changing its nostril size (the elephant has at least 25 different calls, each with a special meaning.)

The elephant's trunk is truly a marvelous tool. It can measure seven feet long and weigh as much as 400 pounds. The trunk alone has more than 100,000 independently controlled muscles; humans have only 639 muscles. The elephant's multi-purpose trunk is far too complicated to have come about by accident and chance.

I am the LORD who made all things.

~ Isaiah 44:24

How does the enormous elephant lower its body temperature? When warm-blooded animals digest their food, they generate heat. Large animals generate lots of heat. Just like a car engine, there needs to be a cooling mechanism; otherwise, the heat would kill the animal – especially in hot climates.

The elephant is a perfect example. It is a large animal that lives in hot climates. As an engineer, how would you design this creature to overcome the heat problem? God has designed the elephant with huge ears. The ears of an elephant weigh about 100 pounds



each. Each ear has a multitude of small blood vessels that can carry enormous amounts of blood through the thin ear membrane. By changing how closely the ears are held to the body or moving air around the ears by flapping them, the elephant can control how much the blood is cooled before it returns to the rest of the body. You could say that elephants have radiators in their ears!

Would you believe that a car's radiator happened by accident or chance? Of course not! So why tolerate you or your children being taught that the even more complex radiator found in the ears of an elephant is the result of chance mutational changes? By the time God had finished His work of creation, He had solved millions of engineering problems just like this one. God displays His great power in the things He has made.

**For the invisible things of him from the creation of the world are clearly seen, being understood by the things that are made...**

**~ Romans 1:20**

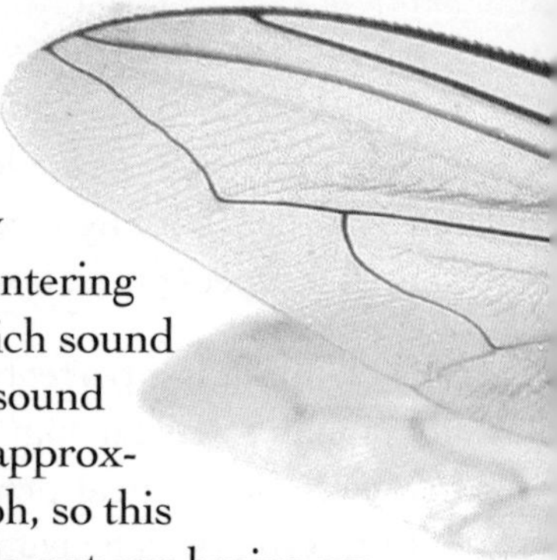

The common fly may have a hearing ability capable of revolutionizing hearing aids for humans. Human hearing is possible because of our ability to sense tiny differences in air pressure waves (sound) entering our ears. We detect the direction from which sound is coming based on the time difference for sound to reach our ears (which are separated by approximately 6 inches). Sound travels at 768 mph, so this difference can be as small as 0.0004 seconds, yet our brains are



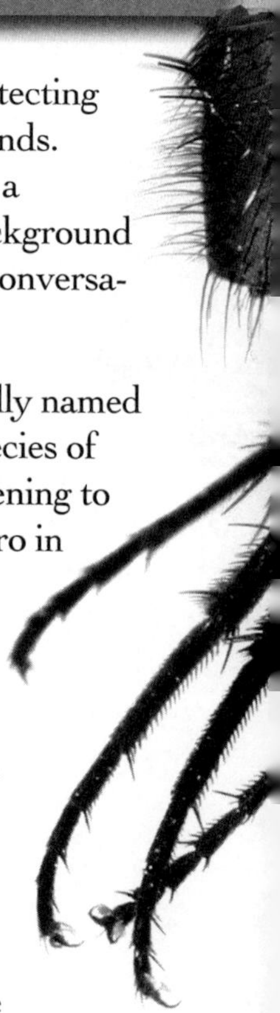

capable of interpreting this difference, immediately detecting the source of the sound and tuning out interfering sounds. This allows people to hear individual conversations in a crowded, noisy room. Yet hearing aids amplify all background noises, making it very difficult to perceive individual conversations.

Enter the hearing ability of the common fly scientifically named Ormia ochracea. This fly lays its eggs on a certain species of cricket. The fly locates the crickets in the dark by listening to their singing. What's remarkable is that the fly can zero in on the singing cricket. How does the fly hear? The distance between the fly's eardrums is minuscule, and it has been found that the eardrums are actually connected to each other. Thus, this fly should not be able to locate the cricket because the fly's eardrums are connected – so there cannot be any discernable difference in sound-arrival-timing. Some alternative, yet-to-be-discovered mechanism for sound direction is apparently waiting to be discovered. This has the potential to revolutionize hearing aid design. We have much to learn from God's design, even the design of a fly's ear.

***...great things doeth he, which we cannot comprehend.***
***~ Job 37:5***

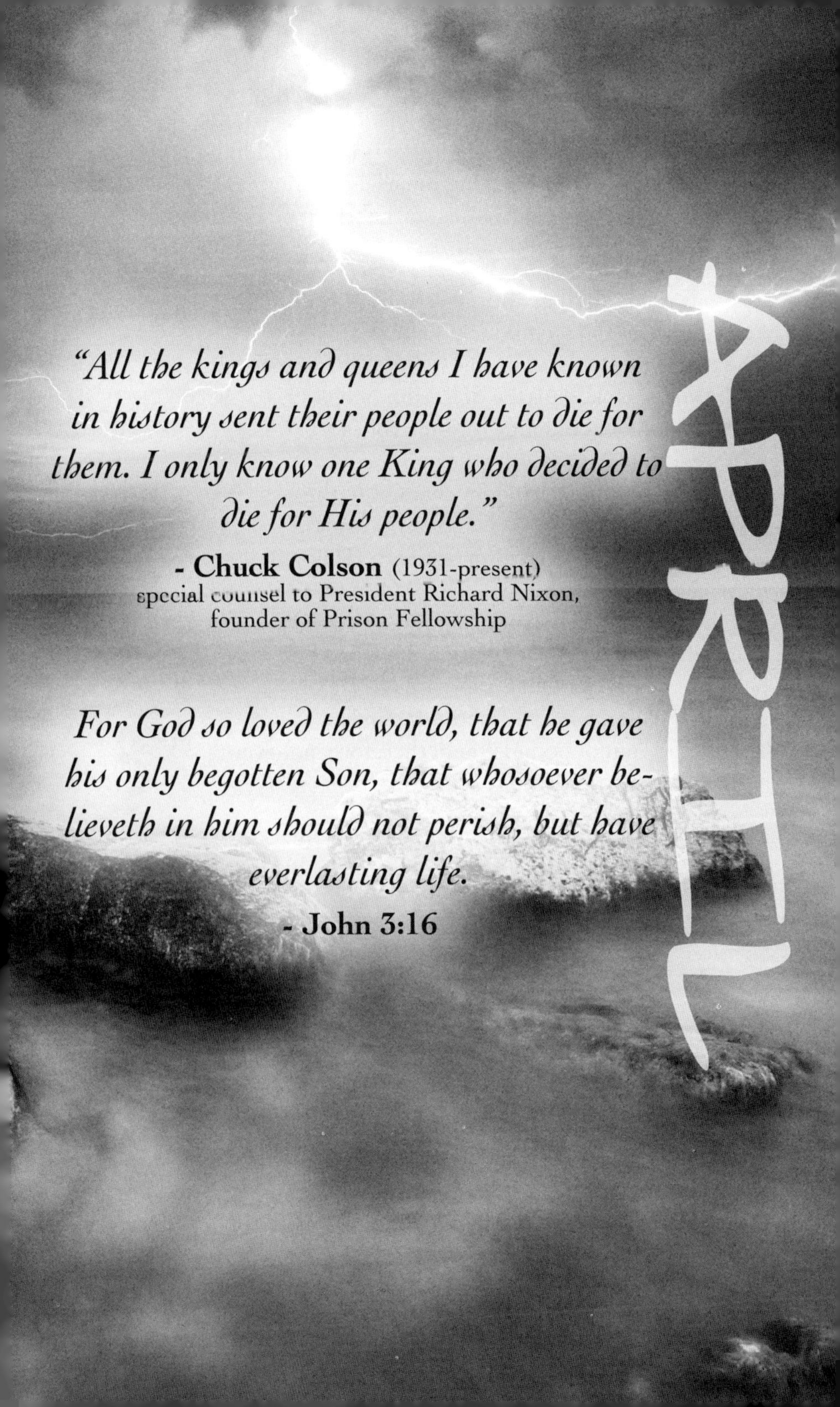
APRIL
"All the kings and queens I have known in history sent their people out to die for them. I only know one King who decided to die for His people."
- Chuck Colson (1931-present)
special counsel to President Richard Nixon,
founder of Prison Fellowship
For God so loved the world, that he gave his only begotten Son, that whosoever believeth in him should not perish, but have everlasting life.
- John 3:16

The primary reason the man-on-the-street doubts the theory of evolution is because he doesn't see new animals, features, or functions appearing around him. Nor does he see books writing themselves, computer codes spontaneously improving, or half-formed creatures appearing on nature shows.

There is no doubt that natural selection works, favoring those changes which have an advantage. The unsolved elephant-in-the-room problem for evolution is explaining the **direction** of changes. The development of man-from-microbes requires an enormous increase in information content. Yet, all random changes to the genome (DNA) information content result in a decrease in functional information.[1] This information degeneration is the "trade secret" of geneticists. Yet, they must ignore this reality in order to believe that evolution developed us because the only possible alternative is supernatural creation by God – and this concept is considered "out of bounds" and, therefore, not even allowed to be considered.

Both scientists and non-scientists are actually observing the same thing – life degenerates, but new,



more complex forms of life are not appearing. The theory of evolution simply does not explain where the diverse forms of life upon our planet came from. Life does not and cannot increase in complexity by itself. God had to have made it, exactly like He told us in His Word.

**The fool says in his heart, "There is no God."**

**~ Psalms 53:1**

The belief in an old Earth is like rat poison. It is 99.9% harmless, attractive and filling. Yet the 0.1% poison ultimately kills. The old Earth position will always be attractive because it agrees with what is commonly taught in schools, museums, and media. If millions of



years are reality, then the fossils in the rock layers testify to a Creator who made death, disease, and bloodshed from the very beginning of creation – long before mankind sinned. Every time someone accepts billions of years of Earth history, they are making God the author and cause of death, disease, extinction, and bloodshed. If a person remains logical and follows this belief to its natural conclusion, it will ultimately destroy the belief in the Word of God, the reality of biblical history.

An old Earth is the absolute non-negotiable core of evolutionism. Evolutionists will give up everything except ancient ages because it was the promotion of enormous time periods that started evolutionary thought. Millions of years are the foundation upon which their entire belief system rests. Evolutionary thought is out to destroy the belief in God's Word, similar to how Satan cast doubt on God's trustworthiness in the Garden of Eden, "Did God really say...?" (Genesis 3:1b)

In the name of trying to make the Bible more credible to the world, old-Earth creation believers ultimately destroy its credibility. Many old-Earth creationists are our brothers in Christ, and we must treat them with respect even though they are seriously wrong on this issue. But they are doing far more damage than good by holding onto an ancient Earth fallacy. Ultimately, those who have accepted the sacrifice of the Lord Jesus Christ for their sins will be in heaven. But, I would not like to be in their place as they stand before their Maker and try to explain why they discarded the Lord's clear, straightforward statements in Scripture, accepted the fallible ever-changing opinions of man, and then, taught others to do the same.

***But whoso shall offend one of these little ones which believe in me, it were better for him that a millstone were hanged about his neck, and that he were drowned in the depth of the sea.***

***~ Matthew 18:6***

The bamboo plant takes five years to mature, showing very little "above ground" activity during the first 2-4 years. Meanwhile, an extensive root system is developing underground. From all appearances, the plant is accomplishing little.

After years of seeming insignificant, the bamboo plant reaps the benefit of its hidden activity – becoming the fastest growing plant on Earth. Nourished by years of unseen activity, at about year five, the bamboo plant sends stalks rocketing skyward at an unbelievable three feet per day.[1]

When serving God, we are seldom allowed to see the true impact. This is by design. If we were serving Him to see the impact of our actions, we would be puffed up – focusing on our accomplishments instead of the One whom we are serving. There would be no faith involved. Like the bamboo plant, most of our growth and accomplishments go unseen. Only the Lord fully understands the impact of our obedience – often generations later. We can be confident that the results of serving God faithfully are grand indeed, even if we seldom witness skyrocketing growth as a result of our obedience.

**I have planted, Apollos watered; but God gave the increase.**

**~1 Corinthians 3:6**

Have you considered sweat? Sure, you know it cools the body, but did you know that scientists have discovered that sweating helps us in another way? Sweat provides food for certain types of bacteria and fungi which live on our skin. Our skin is covered with millions of bacteria at this very moment. Sounds rather creepy, but the opposite is true – the vast majority of bacteria are actually quite helpful.

## INSPIRED BY ANATOMY

APRIL 4TH

These non-harmful bacteria and fungi living in our sweat produce lactic acid, that in turn kills pathogenic (bad) bacteria and fungi. Our skin is designed to protect us against many infections, but our skin is just the first level of protection. God designed our skin to make food, in the form of sweat, so good bacteria and fungi could eat it and make the killer acid needed to destroy pathogenic bacteria and fungi. Without sweat, you would be far more prone to sickness. Sweating keeps you healthy!

The bacteria and fungi need the sweat to live, and you need the bacteria/fungus acid to remain healthy. This is called symbiosis, an arrangement in which two different organisms live together in mutual benefit. If evolution were true, how would these organisms get together at the same time and same place? God knew what the human body would need in order to stay healthy. My spouse would often quip, "You need one good sweat a day." I never knew why, but now I do.

So, get sweaty! God has put together a beautiful design.

*In the sweat of thy face shalt thou eat bread, till thou return unto the ground; for out of it wast thou taken: for dust thou art, and unto dust shalt thou return. ~ Genesis 3:19*

A brightly colored butterfly would seem to be an easy lunch for birds because it flies so slowly. Its wings are not streamlined aerofoil wings like those of birds. Yet, researchers were amazed at the variety of flapping techniques that butterflies displayed in a wind tunnel. There was an entire range of complicated wing movements that generated more lift than simple flapping. It appears that butterflies flutter because they are customizing each wing stroke to match the need. The twists, flaps, slaps and flings are not just erratic wandering, but they are instead designed to keep the birds at bay.

Even though a bird can fly much faster, a butterfly can fly tight, darting patterns which enable them to escape. Have you ever tried to catch a butterfly? It is difficult! So how can butterflies show so much variety in their flight? There is one engineer who knows. He is the One who originally equipped the butterflies with their various flapping techniques. Not only is the butterfly beautifully designed, but it is also designed with the foresight needed to allow these beautiful creatures to survive larger, faster predators.

**That I may publish with the voice of thanksgiving, and tell of all thy wondrous works. ~ Psalm 26:7**

It is common to find dinosaurs in the classic "dead dino posture": the head and neck arched back, tail extended and hind limbs bent. Many dinosaur skeletons have been found in this posture, suggesting extreme agony.

Research has now eliminated the standard idea that this "dead dino posture" was caused by the contraction of the dead animals' tendons as they dried out. Researchers are discovering that animals found in this position died violent, sudden deaths. Animals go into death throes or "opisthotonos posture" shortly before they die because of the shortage of oxygen. This shortage of oxygen causes muscles to go into spasms causing the classic posture while they are gasping for breath. The death throes cause the animal to arch its neck and head, extend its tail and bend its hind legs. What we see preserved in the fossil record are dinosaurs, archaeopteryx and other critters in the midst of their death throes.

What possible event could cause the asphyxiation of so many creatures preserved in the "dead dinosaur posture"? The Flood of



Noah's day devastated the entire Earth. This was a catastrophic, global flood which buried billions of creatures alive. Evidence of this global flood is revealed in the common "dead dinosaur posture." Creatures were rapidly buried and experienced a lack of oxygen, which caused them to go into the throes of death. The next time you are in a museum or see a picture of a fossilized dinosaur or creature, take a look at its posture—see the death throes posture and know it was killed in the Flood of Noah's day.

**All in whose nostrils was the breath of life, of all that was in the dry land, died.**

**- Genesis 7:22**

Humans have always buried their dead. Even Neanderthal and Cro-Magnon Man (people alive shortly after the flood of Noah who found shelter within caves) buried their dead along with valuable tools and artifacts. This tells us two important things about human history.

First, people have always had knowledge that this life is not all there is. From the ancient Egyptians and Chinese to current "stone age" tribes in New Guinea, humans prepare their loved ones for the life after death by burying them along with treasured possessions.

Second, humans could not possibly have been around for 50,000+ years. Suppose:

- For most of these 50,000 years the entire world population averaged only half a million humans.
- These "primitive" humans had an average life span of 50 years.

Even using these extremely conservative numbers, there should be 500 MILLION Stone Age graves out there someplace!

Yet, only a few thousand are known to exist. This is simply because mankind has been on this planet for a few thousand years. Stone age people were simply those few living in caves during the ice age, which followed the flood of Noah. We can understand why so few graves are found when we look to the Bible as the true history of the world.

***All scripture is given by inspiration of God, and is profitable for doctrine...***

***- 2 Timothy 3:16***

# INSPIRED BY BIBLICAL ACCURACY

APRIL 8TH

Every detail of Scripture is recorded for a purpose. All too often we read through many of the stories and historical accounts in the Bible without understanding the significance of the tiny details. This is especially true when we do not understand the cultural customs of the day. One of those customs was how a guest at a banquet handled his napkin during a meal.

Banquets and meals in Middle Eastern culture were an entire evening affair. While eating, guests could come and go from their places, so servants cleaning up the banquet would not want to clear a place if the guest was going to return. Thus a commonly followed custom was to fold the napkin if you planned to return, this signaled to the servants your intention of returning. If you did not intend to return, you would leave the napkin crumpled, so the servants knew they could clear your place.

After Jesus' resurrection, his disciples went into the tomb where they found the flattened burial cloths, which had been wrapped around his body. But the cloth "napkin" that had been wrapped around Jesus' head was not found lying in a crumpled, flattened pile. It had been neatly folded and laid aside, clearly signaling the widely known custom of the day, i.e., Christ was letting us all know – His intention to return!

**Then cometh Simon Peter following him, and went into the sepulchre, and seeth the linen clothes lie, and the napkin, that was about his head, not lying with the linen clothes, but wrapped together in a place by itself.**

**- John 20:6,7**

If evolution is true, it must start by explaining the existence of the only planet upon which life is known to exist (Earth). Almost every astronomy textbook, natural history museum, and planetarium show promote the Nebular Theory as the best explanation for why the Earth exists. The Nebular Theory is the idea that the Earth formed from a mass of swirling dust particles that were circling our sun billions of years ago. These dust particles slowly came together by gravitational forces to form larger and larger



bodies until all of the planets, including Earth, had formed. This idea was first proposed by French atheist Pierre De Laplace in the early 1800's but was almost immediately shown to be totally impossible by mathematical genius James Maxwell.[1]

Maxwell proved that there were two major defects in the Nebular Theory. First, our sun is rotating much too slowly to have been formed from a cloud of swirling gas. Second, clouds of particles do not condense to form planets. This violates the laws of physics– they will always disperse. It is extremely revealing that Maxwell's mathematical proof has never been refuted, yet the Nebular Hypothesis is still taught in astronomy textbooks as the best explanation for the formation of our solar system.[2]

The nebular planet formation model is still promoted not because it explains the origin of our solar system but because evolutionists reject the possibility that the sun and planets were created. Thus the Nebular model, even though it does not work, is simply the *story-with-the-least-problems* used by evolutionists in their attempt to explain the existence of our solar system without God.

**He stretcheth out the north over the empty place,**
**and hangeth the earth upon nothing. - Job 26:7**

Sparrows make flight look easy. As a sparrow darts among forest trees, it never touches the brances or leaves. That's because its feathers are continually adjusting to the surroundings. Side winds are no problem; the feathers are instantly adjusted. Sparrows are made to fly. For birds to fly, they need to be specially designed.

- Bones need to be light weight, so they are hollow and filled with air. If they were filled with marrow, they would be too heavy, and flying would be difficult or impossible.
- Birds have a very short digestive tract. Nutrition moves in and out as quickly as possible, often while flying. Extra weight would hold the bird down.
- Birds have no bladder because that would add unnecessary weight. Birds concentrate their urine. Uric acid needs very little water to be eliminated. We often observe this white paste on our cars. The birds are constantly unloading extra weight.
- Birds do not give birth to live young. A flying pregnant bird would weigh too much. Instead, birds lay eggs.
- Their lungs are a one-way system that circulates air almost throughout the entire body using large air sacs and hollow bones. The air is exchanged in one breath. This system allows maximum oxygen absorption and maximum carbon dioxide elimination.

## INSPIRED BY DESIGN

APRIL 10TH

These are but a few of the design features that birds needed right from the beginning in order to fly. If one were missing, flight would be impossible.

"Does the hawk fly by thy wisdom...?" ~ Job 39:26

Granite is well known for crystals of pink, cream, and black. The black flakes are a shiny mineral called biotite. To the unaided eye not much can be seen, but placed under a microscope, tiny zircons (another mineral crystal) can be seen. These zircon crystals often contain halos of dark rings, resembling targets. These many rings or halos are formed by radioactive decay inside the zircon.

Uranium decay produces a halo with 8 rings. The last 3 rings are made by an element called polonium. Polonium decays so rapidly that it is rarely found. Polonium is generated when uranium decays; so traces of polonium are often associated with uranium. So, it was with great surprise that polonium was found to exist **alone.** How was that possible? As the uranium was decaying, hot water flowing inside the granite carried the polonium a short distance. The polonium then formed its own halos as it decayed. For the polonium halos to be preserved in rock meant the granite had to cool quickly, locking these halos into the stone. It is often claimed that granites cooled slowly over millions of years. If this were true, no polonium halos would exist in granites today. Yet, we find an abundance of polonium halos in granites all around the world.

## April 11th — Physics Inspired By

Granites from Land's End in Cornwall, England contain polonium halos in practically every black biotite flake. The same is true with the Strathbogie Granites of Victoria, Australia. The existence of polonium halos in granites means that the granites had to cool and solidify in just 6-10 days. Polonium halos provided proof that granites cooled and solidified quickly; therefore, the rock layers of the Earth cannot be millions of years old. Granite rocks have revealed that the Earth is young, just as God's Word declares in the historical narratives in Genesis.

**Truth shall spring out of the earth;
and righteousness shall look down from heaven.
~Psalm 85:11**

Bird feathers are a marvel of engineering with an ingenious system of barbs and barbules – creating a high-strength, air-capture surface with an extremely low weight. When examining a feather, one will notice the barb coming off the main stem. On either side of the barb are tiny barbules. These barbules have hooks on one side and ridges on the other side. These hooks and ridges connect together, acting like Velcro™. These hooks and ridges result in a sliding joint, allowing the feather to be flexible and yet remain intact in order to capture the force of the wind. However, this delicate system of hooks and ridges can only work if there is oil or powder to lubricate the sliding joint. Hawks, owls, pigeons and some parrots have specialized down that turns into powder and acts as a lubricant.

Other birds turn their necks 180 degrees to get oil on their bills from a preening gland (uropygial gland) which secretes oil at the base of their spines. This oil is then spread throughout the feathers. Without this oil, the feathers would become frayed and useless within days. Birds spend hours preening their feathers. For the birds to reach this preening gland, multiple neck bones are needed. Humans have only seven bones in their neck while birds have from 13-25 bones.

A feather without the intricate barbules/hook sliding joint would be useless – resulting in no flight. A bird with fewer neck bones would not be able to reach its oil gland – resulting in frayed feathers and no flight. If the bird could reach that preening gland but found no oil, he still would have frayed feathers – again resulting in no flight.

The design of the feather, the numerous neck bones, and the preening gland had to be present from the beginning for flight to take place. It took mankind almost 6,000 years to achieve flight –

## INSPIRED BY DESIGN — April 12th

using the bird's wing design in modern airplanes. God in his wisdom designed the bird to fly and gave birds all the equipment needed to achieve this marvel of nature.

**All things were made by him;**
**and without him was not any thing made that was made.**
**- John 1:3**

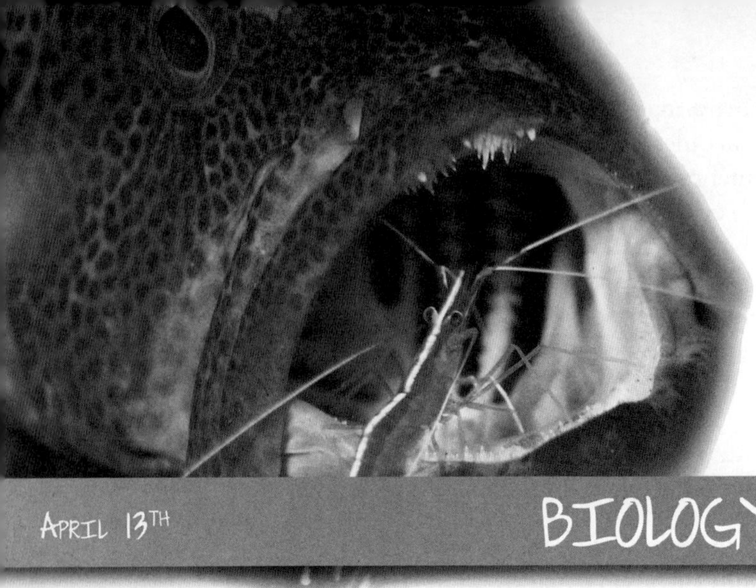

April 13th

# BIOLOGY INSPIRED BY

As fish roam the ocean seeking smaller fish and shrimp to eat, their mouths begin to accumulate food particles. These fish need a toothbrush! So off to the cleaning station they go. Cleaning stations are often set up by the very shrimp these fish like to eat. Fish will even line up and wait their turn to have their teeth cleaned. When its turn comes, the fish will open his mouth wide, and the cleaner shrimp will scuttle inside these jaws of death. The cleaner shrimp will eat the bits of food from between the fish's teeth. Imagine shrimp crawling around on the fish's tongue and picking off parasites and food particles. The fish will even hold its gill chambers open, so the shrimp can crawl around picking off parasites. At the end of the cleaning you would think, "Clean teeth, free meal". Snap. But no, when the cleaning is finished, the big fish lets the cleaner shrimp back out.

Who negotiated this truce? Who was the first brave shrimp? What if he was wrong and was eaten? How does evolution explain this symbiotic relationship? The fish and shrimp work with instincts. Where there are instincts, there must be an instinct maker. That instinct maker is God.

**Let the heaven and earth praise him, the seas,**

**and every thing that moveth therein.**

**~ Psalm 69:34**

The Puritans came to America with few physical possessions but a rich spiritual heritage. As they died, they passed along both the



spiritual and the physical treasures to their children, but all too often, over time, Puritan-established congregations strayed from their spiritual legacy and drifted away from biblical basics. This situation always led to a spiritual showdown with those wishing to stay true to biblical truth. Inevitably, the minority from the Bible believing group was forced to leave the apostate church and move down the road. They would start a new denomination with the comment, "They kept the furniture, but we kept the faith."

What about the spiritual and physical legacy of science? Essentially, all of the founders of modern science had strong beliefs in God and the Bible. Brilliant men such as Sir Isaac Newton, Johann Kepler, Robert Boyle, William Herschel, Michael Faraday, Jedidiah Morse, Louis Pasteur, Matthew Maury, Lord Kelvin, Alexander Graham Bell, George Washington Carver and others saw no conflict between science and the Bible. They had no problem developing the laws of science while believing in God's Word. We have all inherited the modern marvels and technological fruits of this science legacy.

But, today's Bible believing scientists are now locked out of universities, science organizations, science forums, and museum presentations. Naturalism and atheism have absconded with the "furniture" of these great men (universities, government funding), but it is Bible-believing Christian scientists who have inherited the faith. In the end, truth will triumph, but inventions and discoveries will taper off because the scientific majority are no longer "thinking God's thoughts after Him" (a famous quote of Sir Isaac Newton). We must continue, as did the founders of modern science, to understand and control nature's process for God's glory and mankind's good.

**For the earth is the Lord's, and the fulness thereof.**

**-1 Corinthians 10:26**

Fire and forests do not normally mix, yet several species of pine trees seem designed to thrive after forest fires.

- The mature Ponderosa pine bark looks like thick plates or puzzle pieces. When heated during a forest fire, these plates pop off. The tree actually sheds the flames! As the Ponderosa grows, the lower limbs self-prune, thus preventing a ground fire from climbing up to the crown of the tree.
- The Jack pine grows in portions of Canada and the Great Lakes states. The Jack pine doesn't drop all of its ripened seeds. In fact, most of the seeds are protected safely within closed cones for years. When a fire occurs, these seeds are protected from the intense heat. However, this same heat opens the cone, releasing the seeds onto the ground after the fire passes. The fire has prepared the ground for the seedlings to grow by burning up the existing vegetation.
- The Longleaf pine, or the legendary southern yellow pine, is designed to survive a fire in a unique way. The seedling, if top-killed by a fire, sprouts from the root again. Once the terminal bud develops, the moist, dense tuft of needles protects it. If there is a fire, the tuft of needles burns and water is vaporized from the needle tips. The steam produced extinguishes the fire. As the yellow pine ages, the bark becomes thick, which insulates the inside of the tree from heat. If the bark starts to burn, the scaly bark pops off. This tree also sheds the fire!

These pine trees did not get together and talk about the threat of forest fires and what to do about their future survival. Had these mechanisms not been in place before fires happened, the trees could never have survived in the first place. Pine trees and pine cones do not have the ability to think. God, in His wisdom, designed these pines to survive a fire in this manner.

*Every tree that bringeth not forth good fruit is hewn down, and cast into the fire.*

*- Matthew 7:19*

There are an overwhelming number of archeological confirmations of the Bible's accuracy. Here are just a few:

- **Crucifixion of Jesus:** In the gospels, Jesus is nailed on a cross. Critics have questioned if nails were really used - as opposed to tying the victim to the cross. Recently, an excavation revealed a crucified victim. The 24-28 year old man was found in a tomb near Jerusalem with a 7 ½ inch long nail through his feet. His crucifixion was dated at A.D. 42 (Jesus was crucified about A.D. 30). Furthermore, the calf bones were brutally fractured, clearly from a single blow. The gospel of John states, "The soldiers, therefore, came and broke the legs of the first man who had been crucified with Jesus and then those of the other. But when they came to Jesus and found that he was already dead, they did not break his legs". Normally the Romans left the crucified person undisturbed to die slowly of asphyxiation. However, Jewish tradition required burial on the day of execution. Therefore, the executioner would break the legs in order to hasten death.
- **The Census:** We now know a Roman Census took place every 14 years – beginning with Augustus. In Luke, Joseph and Mary had to go to Bethlehem for a census.
- **Pontius Pilate:** Pilate was identified in the Bible as governor of Judea at the time of Christ's crucifixion. Archeologists dug up the site of ancient Caesarea and found this inscription, "Pontius

## INSPIRED BY BIBLICAL ACCURACY APRIL 16TH

Pilate, the Prefect of Judea, has dedicated to the people of Caesarea a temple in honor of Tiberius."
- **Sodom and Gomorrah:** For many years it was thought that references to these cities were merely a moral lesson and were not actual historical places. We now know that they were real cities. In the excavation of Ebla, a city in northern Syria, some 20,000 tablets have been found. Inscriptions on some refer to the cities of Sodom and Gomorrah as trading partners of Ebla.

Archeology again and again, proves the Bible is totally trustworthy.

**But whoso hearkeneth unto me shall dwell safely,**

**and shall be quiet from fear of evil.**

**~ Proverbs 1:33**

Work is considered by many as a necessary evil, a curse, something to be avoided at all costs. Yet, work was given to mankind *before* people rebelled against their maker, and God allowed death, disease, and *hard* toil to enter into all of creation. Work is undoubtedly a blessing from our Maker. Christians should acknowledge and act upon this fact. When Christians truly understand biblical creation, their perceptions of work are transformed from curses to blessings.



If Christians demonstrated Biblical characteristics of a strong work ethic, honesty, creativity, and a positive attitude, they would be in great demand as employees. Every aspect of our lives should reflect these characteristics because we are made in God's image. God worked for six days to make all of creation. God took joy in work. God showed incredible creativity in His work. God then rested on the seventh day as a pattern for us to follow. Thus, He is honored when we do the same.

After finishing the work of creation, God did the most incredible thing of all – he placed mankind in a stewardship/manager position over all that He had made. Every invention, every song, every discovery, every painting, every medical advance, every tool, every social structure – all of mankind's technological and artistic accomplishments – were anticipated and pre-known by God. He gave us the entire world as a playground for discovery and invention. Even mundane work can be rewarding and inventive when we approach it with the attitude of creativity and gratitude.[1]

**And the LORD God took the man, and put him into the garden of Eden to dress it and to keep it.**
**– Genesis 2:15**

It is incredibly damaging to remove meaningful work from people. This can happen in two ways – either work can become intolerably mundane with no particular purpose, or people can be given all that they need for subsistence with no effort required on their part. Both



result in catastrophic consequences, both individually and for society at large.

In a cruel experiment, prisoners were divided into two groups. One group was given the meaningless task of moving dirt back and forth, day after day, with no purpose or end in sight. These P.O.W.'s rapidly died. A second group was worked as hard, or harder, but their tasks had obvious purposes. This group endured the labor with far less loss of life. Meaningless labor or the removal of all hope in life is pure evil.

From rich children given everything without responsibility, to welfare societies that distribute goods independent of effort expended, these practices are as damaging as the meaningless labor in a P.O.W. camp. The Bible makes it abundantly clear that all who are capable of work should work to earn what they need for survival. God is honored when we help the helpless (widows, orphans, and cripples), and God is pleased when we provide opportunities for employment for the jobless, BUT the New Testament makes it very clear that those capable of work must do just that.

Before the fall of creation, work was given to mankind as a gift. The Old Testament states that land owners were commanded to allow the poor to glean (collect) leftover food from their fields – in essence, to work for their survival. This is because work (of any meaningful kind) builds our self esteem, character, perseverance, and appreciation for God.

*For even when we were with you, this we commanded you, that if any would not work, neither should he eat.*
*~ 2 Thess. 3:10*

Have you considered how birds "hear" blizzards coming?

Birds have exceptional hearing but in a lower frequency range than normal human hearing. Birds can actually hear bad weather approaching. I am sure you have noticed that a day or so prior to a major storm, birds are frantically feeding at your feeder. These birds hear the storm coming and prepare for it. Also, birds such as hawks and eagles can actually hear the low frequency sounds of approaching thermal air changes. This is good because their very lives depend on thermal currents in order to soar. Birds can

also detect very minute differences in air pressure. You could say they have barometers in their ears. Birds need to be very aware of the weather and the changes taking place in order to survive.

Other animals also have incredible hearing abilities. Whales communicate in the infrasound range with each other at distances of more than 100 miles. Male elephants also use these frequencies to communicate up to six miles away. God cares for the very life of His creatures and has designed them with the abilities they need to survive.

*Are not two sparrows sold for a farthing? and one of them shall not fall on the ground without your Father. Fear ye not therefore, ye are of more value than many sparrows. ~ Matthew 10:29,31*

If saltwater covered the world only 4,500 years ago during the worldwide flood, why weren't all land surfaces spoiled by the resulting salt? This is a challenge commonly promoted by those who reject any possibility of a global flood upon the Earth because a worldwide flood would have created the Earth's rock layers rapidly, removing the time needed for evolution. Thus any excuse to reject the reality of this flood is promoted.

Covering land with salt water for a few months would not saturate the land with salt because there would be no widespread



driving force to precipitate salt out of solution. The Genesis flood covered **the entire** Earth for months and sediment-carrying waves of water repeatedly washed over many land surfaces around the globe for over a year. Yet, within weeks of the water's retreat into the world's ocean basins, vegetation would have started to reclaim the land. Enormous mats of vegetation would have been both buried within the sediment layers left by this massive catastrophe (creating coal seams) and scattered on top of the newly formed land surfaces. At the conclusion of the flood, the water rushing off the land surfaces would have carved deep, oversized river canyons and vast erosion features. These features are exactly what we find as we observe the geology of our planet. But the salty water would not have spoiled the land for rapidly re-rooting vegetation.

***Or who shut up the sea with doors, when it brake forth, as if it had issued out of the womb?***

***- Job 38:8***

The feather of an owl is not like other birds' feathers. Other birds produce feathers with straight, crisp edges. An owl's flight feathers have fringes at the edges. The owl also has a notch on one of its flight feathers. Both the notch and the fringes provide the owl with a noiseless flight. In fact, an owl is one of nature's stealthiest hunters. Its prey is often unaware it is about to die until it's too late.

Question: What do computers and owls have in common? Computers need fans to keep them cool, yet fans are

noisy. What to do? That was the question posed to a group of computer engineers. One of the engineers noticed an owl flying noiselessly. He saw it flying through the woods, but could not hear a whisper of moving air as it whooshed down upon its prey! The owl was noiseless as it moved air around. Ah ha! Copy the owl feather! Thus, quieter computer cooling fans were created based on the design of the owl's feathers.

*He is thy praise, and he is thy God, that hath done for thee these great and terrible things, which thine eyes have seen.*

*- Deuteronomy 10:21*

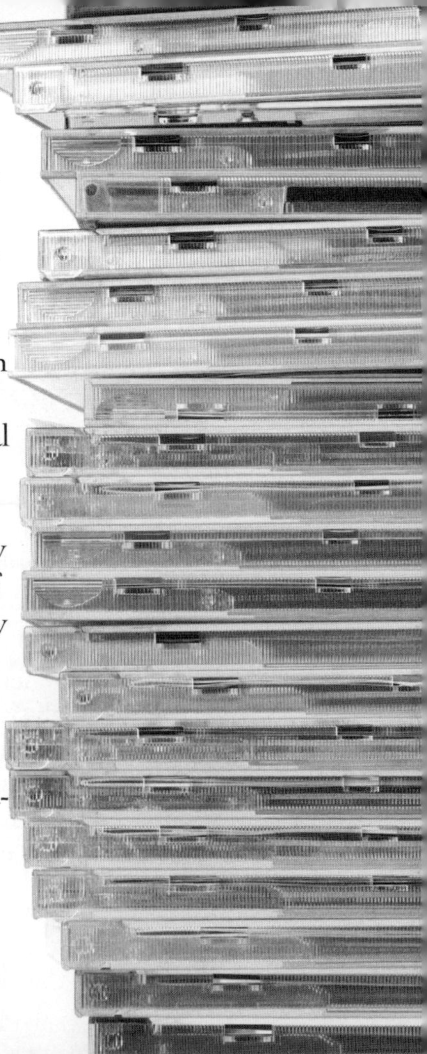

DNA has been called the most mysterious molecule in the universe. The information that it contains cannot be explained by any natural process. The sequence of chemicals on the DNA molecule is used to construct every other molecule in the human body. Yet, neither the structure nor the information contained therein can be reproduced in a laboratory by mixing the four chemical components from which DNA is constructed. All attempts to do so have been as fruitless as trying to develop any of the 22 million books in the Library of Congress by having a monkey randomly doodle symbols on a piece of paper.

One gram (0.0022 lb.) of DNA can hold as much information as one trillion CDs[1]. Stated another way, this information storage system is so efficient that a microscopic speck smaller than a dust particle could hold a copy of every high resolution movie ever produced – with room to spare for every movie made for the next hundred years. As advanced

# INSPIRED BY MICROBIOLOGY

APRIL 22ND

as modern technology has become, mankind's greatest achievements and mighty technological advances are child's play when compared with that which God has produced.

**Oh that men would praise the Lord for his goodness, and for his wonderful works to the children of men!**
**- Psalm 107:31**

Have you thanked God for your kidneys lately? These two fist-sized organs located in your abdominal cavity do the following:

- Filter out waste products from your blood.
- Regulate the blood stream to be 51% water. A 5% fluctuation could cause your blood to be too thin and your blood pressure to drop, or be too thick, so your heart could not pump it.
- Maintain the perfect concentration of salts within your body.

## April 23rd — ANATOMY INSPIRED BY

- Regulate the acid-base equilibrium of your blood.
- Secrete hormones that help maintain blood pressure and make red blood cells.

Every 50 minutes, every blood cell in your body passes through your kidneys. When a person's kidneys fail, he needs kidney dialysis. Several times each week, a dialysis patient must spend hours circulating his blood through this very complex machine. This machine is hundreds of times larger and far less efficient than our kidneys.

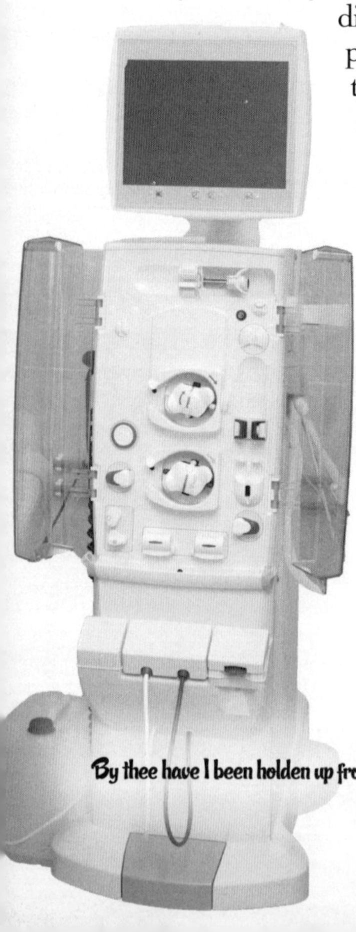

Evolutionists say the kidneys happened by chance - that they were essentially a useful accident. Did the kidney dialysis machine happen by chance? Is the kidney dialysis machine (which was planned and designed) better than these "accidental" kidneys? No, it is not; it can only replicate a small fraction of the hundreds of processes which take place within our body's kidneys! It has been claimed that next to the brain, the kidneys are the most complicated organ in the body. Quite a problem to explain via random mutational changes!

**By thee have I been holden up from the womb: thou art he that took me out of my mother's bowels: my praise shall be continually of thee.**

**~Psalm 71:6**

Mars is a dry, frigid world, yet recent discoveries of enormous erosion patterns suggest the possibility that the planet experienced huge floods in the past. Mars was created on Day 4 of creation week, three days after Earth. As we read in Genesis 1:16, "God made two great lights; the greater light to rule the day, and the lesser light to rule the night: he made the stars also..." The Hebrew word for "stars" refers to any bright object in the sky, which would include such things as meteors, stars and planets. So Mars was made on Day 4.

## INSPIRED BY COSMOLOGY

APRIL 24TH

The search for water on Mars has resulted in finding polar caps made of dry ice or solid carbon dioxide. Even though liquid water has not yet been confirmed, Mars has many features that point to huge floods in its history. Secular scientists even refer to a "Noachian epoch" on Mars. These same scientists deny Earth having had a global flood – even though the Earth's oceans contain enough water to cover the entire surface of our planet 1 ½ miles deep – if the ocean basins were shallower and mountains flattened down!

It is difficult for non-believers to admit their accountability to their Creator, so they deny a global flood on Earth (which has abundant water) while accepting a flood on Mars (which has essentially no water).

**[Scoffers will come] And saying, Where is the promise of his coming?... Whereby the world that then was, being overflowed with water, perished... ~ 2 Peter 3:3-6**

The Bible claims that during Jesus' crucifixion, "the sun stopped shining". Darkness was upon the Earth from 12 noon to 3 pm. Some people claim that the darkness was no supernatural event, only a solar eclipse. Let's examine this idea.

- First, we know Passover was being celebrated at this time, and the date for Passover is based on the moon cycle (that's why the date for Easter varies so much). Passover takes place on a full moon. A solar eclipse could not have taken place during a full moon; the moon would have been in the wrong position. For a solar eclipse to occur, it needs to be a new moon.
- Second, the text says darkness came upon the land from the sixth hour until the ninth hour. No solar eclipse could have lasted that long.
- Third, the text says darkness was over the "whole" Earth. A solar eclipse would have been a very local event. Just as liberal theologians destroy the credibility of Genesis by ignoring that the most common word in Genesis 6-9 is the word "all" (i.e. flood waters covered "all" the hills under the heavens), they are willing to deny the text in an effort to explain away the supernatural.
- Fourth, this event caused the centurion to proclaim, "Truly this was the Son of God" (Matthew 27:54). People had seen many solar eclipses. Although they may have been feared, seeing one during the crucifixion of a man on a cross would hardly cause a Roman soldier to proclaim that that dead man was God on Earth (i.e., the Messiah – the "Son of God").



Those who want to assign natural events to every miracle of the Bible need to ignore the clear, straightforward statements of the text in order to do so. This does far more damage than good because it allows the Bible to mean ANYTHING they want it to mean. If the Bible can mean anything, the Bible means nothing.

**And it was about the sixth hour, and there was a darkness over all the earth until the ninth hour.**

**~ Luke 23:44**

The DNA of all domestic dogs has been shown to be traceable back to a single pair of grey wolves from the past. There are several amazing implications of this discovery. First, it means that the unbelievable variety of modern dogs was present within the DNA code of the original grey wolf before this domestication of modern dogs took place. Thus, the huge variation in dogs – from Chihuahuas to Great Danes, Dalmatians to Dachshunds – has nothing to do with "evolution". All of these varieties are simply a sorting of information already present within the "wolf kind."



Second, it illustrates how unreliable the time frames associated with the appearance of biological life really are. There are ongoing disagreements between geneticists and anthropologists as to when the domestication of modern dogs happened, but all agree that modern dogs must have appeared within the last 10,000 years or so. This is WAY too short of a time period for any significant evolutionary change to have occurred. Again, this confirms that dogs are a result of the sorting of pre-existing information and a recent creation.

God created dogs in a way that allowed for great variety.

**Let the land produce living creatures according to their kind...**

**- Genesis 1:24**

The Redwall Limestone layer in the Grand Canyon is one of the most distinct sediment layers in the mile-deep canyon. Of the nine major layers making up Grand Canyon strata, this 600 foot thick layer covers over 5700 square miles ranging throughout much of Arizona and southern Nevada. The standard geological explanation for the formation of this rock layer required millions of years of settling at the bottom of an ancient sea.

In 2002, Dr. Steve Austin presented to the American Geological Society evidence for a vastly different Redwall formation model. Dr. Austin has a Ph.D. in geology from the University of Pennsylvania but unlike most geoogists trained via our evolutionary presuppositions, he does not eliminate the possibility of a real worldwide flood upon the Earth approximately 4500 years ago when interpreting geological formations. Austin discovered a mass burial of nautiloid fossils near the bottom of the Redwall Limestone deposit. Even more significant, the sugar-cone-shaped fossils were statistically oriented in a single westerly flow direction throughout the deposit.



If this deposit was really a result of slow accumulation over huge time periods, this orientation would make no sense. However, if the entire 600 foot thick deposit is the result of a huge flood flow event which trapped a large group of nautiloids (estimated at over one billion individual creatures), then the arrangement and characteristics of the rock layer are exactly what one would expect to find.

The implications of this find are stunning. If one of the largest and most visible limestone layers in the Grand Canyon is the result of rapid deposition and flood flow processes, then the entire sedimentary record throughout the canyon, from the top rim to the basement rock, was also the result of rapid flood accumulation. The Genesis flood was a powerful, Earth-reshaping flood.

**Whatsoever the LORD pleased, that did he in heaven, and in earth, in the seas, and all deep places.**

**~ Psalm 135:6**

Our month is based on the moon, our year on the sun, but what about a week? The seven day week is based on God's creation

## INSPIRED BY HISTORY

APRIL 28TH

week. Atheist dictator Stalin tried to erase God from the Soviet Union by instituting a 10 day week. People would work nine days and then rest the tenth day. It was not successful because people's bodies could not stand the strain of working nine days with only the tenth day of rest. Others have also tried to eradicate the seven day week, but to no avail. The pattern of a seven day week is a reminder of God creating everything in six days and resting on the seventh day.

We also find God reminding us of this pattern in Exodus 20:8-10, "Remember the Sabbath day by keeping it holy. Six days you shall labor and do all your work, but the seventh day is a Sabbath to the Lord your God, on it you shall not do any work… For in six days the LORD made the heavens and the Earth, the sea and all that is in them, but he rested on the seventh day." Here we see the pattern for a week repeated again; work six days and rest the seventh day.

The God of all creation could have created everything in a split second, but He did not. This all powerful God knew what we needed; He knew what our bodies could handle. From the very beginning of time, He set the pattern for a seven day week – for our benefit.

**Then he said unto them "The sabbath was made for man, not man for the sabbath." ~ Mark 2:27**

Myles Willard is a former high school science teacher, an award winning nature photographer, and an avid bird watcher. Being a systematic observer of nature, Myles noticed that during the fall of each year a large number and variety of warblers stopped to rest in a cedar tree outside of his Michigan kitchen window. He decided to document both the variety and number of warblers migrating south. For the next 18 years, Willard compiled an impressive documentation of over 1,500 individual warbler sightings from over 20 different species.



It was upon graphing the results of these sightings that Myles made a shocking observation. He expected to see a typical bell-shaped distribution of bird sightings with the peak happening somewhere in mid-September (depending on the particular species). When he aligned the various distributions so that the peak of the migration for each individual bird species corresponded, Willard discovered a bell-shaped distribution with statistically significant dips at a 7-day period. Instead of a smooth bell curve, his data showed a bell distribution with drops (like missing teeth in a saw) every 7 days. Apparently birds have a built-in instinct to work hard on their migration for six days (out of seven) and then hold up for one day of rest!

There is no naturalistic explanation for this very significant biblical cycle. But, a rest period every seventh day makes perfect sense if God did indeed create everything in six literal days and institute a day of rest on the seventh day. It seems quite plausible that we should find some remnant of this cycle in the natural pattern of some of the creatures that God made. It would seem that Myles Willard, science teacher, nature photographer, and bird watcher, has found and documented such a pattern.

***Doth the hawk fly by thy wisdom, and stretch her wings toward the south?***

***~ Job 39:26***

Fibonacci numbers are one of many evidences for God's existence. So what are Fibonacci numbers? They are a patterned sequence that can be found everywhere throughout nature: 0,1,1,2,3,5,8,13,21,34...and so on. Every number in the pattern is the sum of the two previous numbers: 0 +1=1, 1+1=2, 1+2=3, 2+3=5, and so on.

One common example is the arrangement of flower petals. The Black-eyed Susan has 21 petals, a Fibonacci number. The Shasta daisy has 34 petals, again a Fibonacci number. The arrangement of leaves around a plant stem display Fibonacci numbers. A 3-sided banana displays a Fibonacci number. The cross section of most bell peppers reveals 3 chambers; 3 is a Fibonacci number. Cut an apple and the seed pod will reveal 5 points on a star; 5 is a Fibonacci number. We see the number five again in the number of arms on a starfish. If you count the spirals on a pinecone, you will find they go in two directions – 13 spirals open to the left, and 8 spirals open to the right – both 13 and 8 are Fibonacci numbers. In fact, there are so many Fibonacci numbers in nature that there is a publication dedicated to it, the Fibonacci Quarterly! As Galileo stated, "Mathematics is the alphabet with which God has written the universe."

There are patterns all around us. When we see patterns, we know there is a pattern maker. The next time you take a walk through that flower garden, don't just stop and smell the roses, but…stop and count the petals...see the pattern and praise the Pattern Maker.

*Who can speak of the mighty deeds of the Lord,*
*Or can show forth all His praise? - Psalm 106:2*

# MAY

*"Not everything that counts can be counted and not everything that can be counted, counts."*

**- Albert Einstein** (1879-1955)
German-born American Physicist

*For by him were all things created, that are in heaven, and that are in earth, visible and invisible, whether they be thrones, or dominions, or principalities, or powers: all things were created by him, and for him.*

**- Colossians 1:16**

Scientists have discovered a bacterium called the Rhodoferax ferrireducens that is capable of producing an electric current as it transforms natural sugars into other complex biological products. It does this by transferring electrons to iron and other minerals in the surrounding water. Found in marine sediments, these bacteria are being explored for use in electric producing fuel cells. This little microorganism does what no human engineer has been able to do. It produces electricity while feeding on sugars such as glucose or fructose with an astounding 80% efficiency. These unbelievable bacterium are twice as efficient at making electricity compared to coal burning power plants.

Evolution says this happened by accident and chance. Does making electricity for our homes happen by accident? No, it involves hundreds of processes. So too with this bacterium; it involves hundreds of chemical processes that scientists do not fully comprehend even today. These chemical processes would have to be in the right order with none missing. Within our electrical generating plants, every part has to be present for them to work and many lines in place to bring the current to each home. If any part breaks or is missing, the plant shuts down, and there is no electricity. When we make electricity it does not happen by accident or chance. In the same way, this bacterium's ability to make electricity did not happen by accident but was designed by its all powerful designer, God. God's creativity simply knows no bounds.

## May 1st — MICROBIOLOGY INSPIRED BY

**And God said, "Let the waters bring forth abundantly the moving creature that hath life, and fowl that may fly above the Earth in the open firmament of heaven." And God created great whales, and every living creature that moveth, which the waters brought forth abundantly, after their kind, and every winged fowl after his kind... – Genesis 1:20–21**

Mankind has entire libraries filled with books about the stars. Yet, we cannot adequately explain how a single one could form! The Bible is not a textbook on astronomy, but when it speaks of scientific subjects, it is 100% accurate. Genesis 1:16 states, "God



made two great lights – the greater light to govern the day and the lesser light to govern the night." How did the writer know that the sun was larger than the moon? Ordinary observation would cause us to think the opposite or that they're the same size. We observe the giant harvest moon on the skyline, and yet we have never observed the sun to be that large. Many ancient peoples thought the moon was far greater than the sun. Yet here we see God telling us the sun is bigger than the moon. Of course, today we know this to be true.

With astonishing brevity, God speaks of the stars with only five words. In contrast, nearly 50 chapters of the Bible discuss the construction and significance of the Tabernacle. Hmmm, fifty chapters about the Tabernacle, five words about the estimated $10^{22}$ stars in the universe. The Bible certainly has a different perspective than ours. The Bible's central theme is redemption. To create stars was child's play for God; He only had to speak. Ah, but for God to redeem us, God Himself had to suffer. That's the Bible's perspective.

If man had written the Bible without the controlling influence of the Holy Spirit, the book would have been quite different. Entire chapters would have been written about the billions of stars in our galaxy and the 100 million other galaxies. There are more stars in space than there are grains of sand on all the seashores. Yet, God dismisses all of it with five words. God is more interested in people and their souls…than all the stars in the universe.

***…he made the stars also. ~ Genesis 1:16***

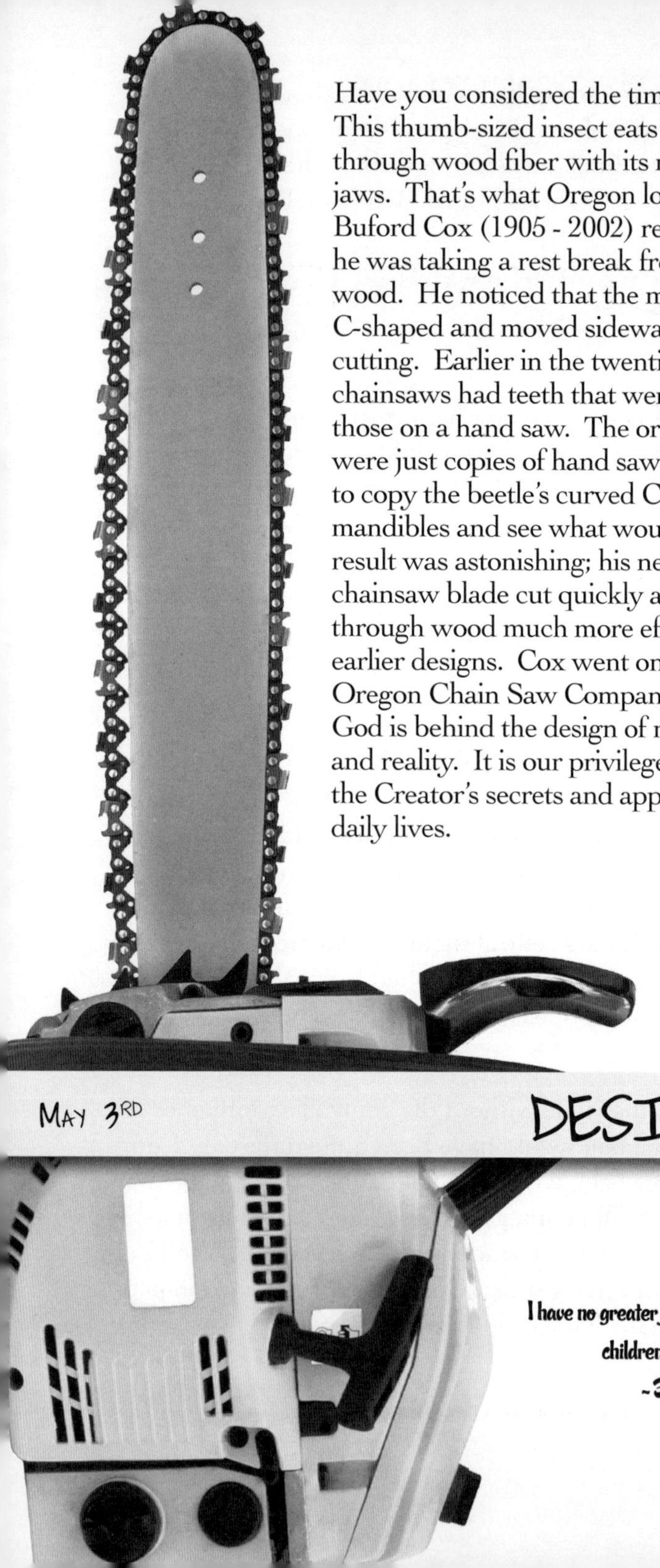

May 3rd

# INSPIRED BY DESIGN

Have you considered the timber beetle larva? This thumb-sized insect eats trees by slicing through wood fiber with its mandibles, or jaws. That's what Oregon logger Joseph Buford Cox (1905 - 2002) realized when he was taking a rest break from chopping wood. He noticed that the mandibles were C-shaped and moved sideways, like scissors cutting. Earlier in the twentieth century, chainsaws had teeth that were pointed like those on a hand saw. The original chainsaws were just copies of hand saws. Cox decided to copy the beetle's curved C-shaped mandibles and see what would happen. The result was astonishing; his newly designed chainsaw blade cut quickly and cleanly through wood much more efficiently than earlier designs. Cox went on to establish the Oregon Chain Saw Company. The fact that God is behind the design of nature is truth and reality. It is our privilege to search out the Creator's secrets and apply them to our daily lives.

**I have no greater joy than to hear that my children walk in truth.**
**- 3 John 4**

INSPIRED BY BIOLOGY

MAY 4TH

Why don't the chickadee's feet freeze when it is minus 30°F? How can ducks spend the whole day in icy water and not get frostbite? It seems that birds are never bothered by cold feet. If we were to run outside in the winter with bare feet, we would have frostbitten toes. So why don't birds need socks?

God has designed birds' feet with a unique heating system. Birds have a network of arteries down to their feet. Interwoven in these arteries are the returning veins. This network is called a "wonder net." The bird's hot arterial blood (106°F) intertwines with its cool, returning vein blood. As a result, the bird's feet have a constant pool of warm blood keeping them from becoming dangerously cold, and the blood returning to the main part of the body is pre-heated.

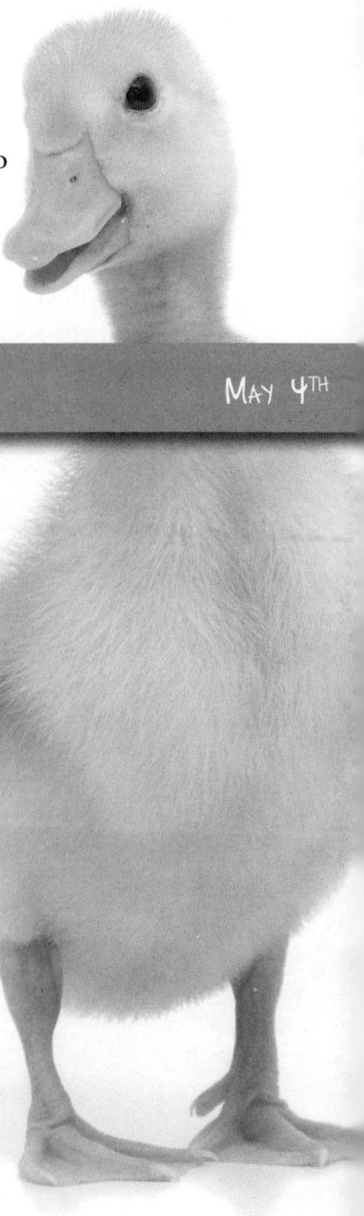

If evolution is true, how many feet had to freeze off before the bird got it right? God in His wisdom designed the bird to survive cold winters. So the next time you see a chickadee at your winter birdfeeder, be amazed how God solved the problem of cold feet for the chickadee.

**How beautiful are the feet of them that preach the gospel of peace, and bring glad tidings of good things!**

**~ Romans 10:15**

Grass is a seemingly unimportant plant that is often ignored and taken for granted. Yet, let's examine it. Notice that the stem is round and mostly hollow except at the joint or node. This stem lies in the sheath that is open part of the way down. The unique aspect of all grasses is their ability to grow from the back of the sheath. Even when most of the grass plant gets eaten or mowed, the grass is able to grow back. This differs greatly from other plants because they grow from the tips of their stems. Grasses are unique in that they continue to grow even if chewed off or mowed. In fact, over one third of the land surface of the world is covered by some variety of grass. Grasses provide food for animals and, thus, man. They don't have to be replanted or sown; they just continue to grow. This seemingly unimportant plant is an enormous blessing to our world. Through it we can see God's provision for our care.

The grass withereth, the flower fadeth: but the word of our God shall stand for ever.." ~ Isaiah 40:8

The weather patterns of North and South America have been periodically affected by El Niños for thousands of years. The ancient farmers of the Andean mountains were very much aware of them. These farmers could predict if an El Niño was present and hold off planting their potatoes, waiting for the rains to come. How would they know to delay their planting? In the early dawn skies between June 13 and 24, these farmers would search the skies for signs of increased moisture. An El Niño, with its warm waters, would produce more moisture that would build up in the atmosphere. While invisible, the extra moisture in the sky would cause some dimmer stars to disappear.

INSPIRED BY BIBLICAL ACCURACY — MAY 6TH

Andean farmers would look to the Pleiades or "Seven Sisters". If all seven stars were shining clearly morning after morning, the Andean farmers knew that no El Niño was forming. However, if day after day they saw a dimming of the stars, they would delay planting four to six weeks because they knew rains would be late. El Niños have been known for centuries, and ancient peoples knew they would affect the weather. They used the stars to help them predict this change in weather. The Bible teaches us that the sun, moon, and stars were made to give us signs about the seasons.

**And God said, Let there be lights in the firmament of the heaven to divide the day from the night; and let them be for signs, and for seasons, and for days, and years. ~ Genesis 1:14**

How did animals spread across the world after they left Noah's Ark? How did the kangaroos get from the Middle East to



Australia? The silly answer would be that they hopped. But really, how did they get to Australia? Isn't there a lot of water between Asia and Australia?

After the Flood, the animals left Noah's Ark, spreading out and filling the Earth. Flying creatures could fly, but what about those that walked? Many could have floated on vast floating log debris left over from the Flood. Others could have been taken by people. One study has suggested that Australian dingoes descended from a single female domesticated dog from Southeast Asia. A third possible explanation is that animals could have crossed on land bridges. One land bridge from Asia to the Americas is over the Bering Straits. For such land bridges to exist, we would need the sea levels to be much lower. After the Flood, during the Ice Age, enormous amounts of water from the oceans would have been locked up in the great ice sheets. This would then lower the sea levels around the world, creating land bridges. It is currently believed ocean levels were 200 – 300 feet lower during the Ice Age – exposing much more land above sea level.[1]

England would have been connected to Europe. Evidence bears this out because we find the remains of many Ice Age animals, such as the reindeer and hippopotamus, in southern England. A land bridge from Asia to Australia would have developed, allowing animals to migrate to Australia. When we view the world from the biblical perspective, there was a catastrophic world wide flood that caused the one and only Ice Age. This would have locked up great quantities of water on land, causing the sea level to drop and creating land bridges for animals and people to fill the Earth – just as God commanded.

**Bring forth with thee every living thing that is with thee, of all flesh, both of fowl, and of cattle, and of every creeping thing that creepeth upon the earth; that they may breed abundantly in the earth, and be fruitful, and multiply upon the earth. - Genesis 8:17**

Many insects live short lives. For example, the fruit fly has a life span of a mere 2 months. In that brief span of time, it must hatch, grow from larvae to adult, mate, and produce the next generation. The mayfly lives longer during development, but only lives 2 days as an adult before dying. This adult insect does not even have functioning mouth parts and is incapable of eating. Why would God create a creature with such a short life span that does not even have the ability to eat food? What advantage could such a short life have? Perhaps God created this creature as a reminder that what matters is not the length of our life, but how it is lived.

## INSPIRED BY BIOLOGY

MAY 8TH

It is possible to spend a lifetime chasing goals only to find them meaningless. Our activities will never bring fulfillment because even as some goal is reached, we feel the emptiness creeping in. As life rapidly passes, we begin the mental gymnastics of focusing on past accomplishments, even though they did not bring lasting fulfillment when first achieved. Life is too often so filled with "seeking" that we forget to enjoy what we have been given.

Life is the process, not the accomplishment. Like the mayfly, the length of our life is a flash in perspective of eternity. So what is God teaching you through your current problems, trials, or issues? What bit of beauty, marvel of creation, or awe of majesty are you missing in a sunset, smile, or snail because you are too busy striving for what you do not yet have?

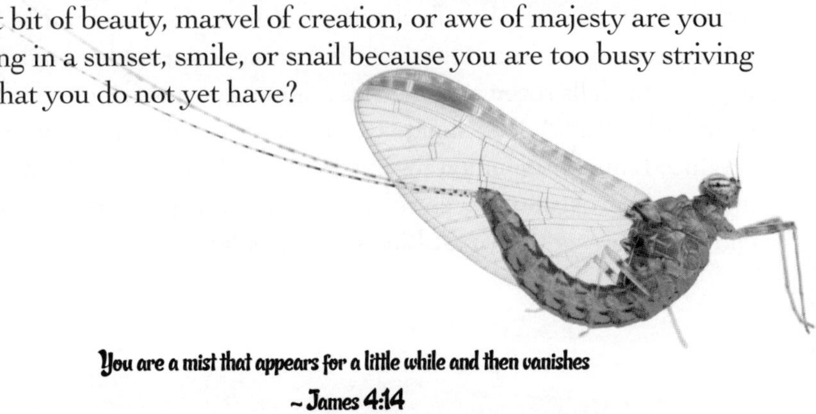

**You are a mist that appears for a little while and then vanishes**

**~ James 4:14**

**Teach us to number our days, that we may apply our hearts unto wisdom.**

**~ Psalm 90:12**

During the Ice Age, deserts and semi-arid regions were well-watered, while 30% of land masses were covered in ice. Geologists studying the Sahara Desert have been surprised to find unfossilized bones of giraffes, buffaloes, rhinoceroses, elephants and other animals along with bones of amphibians, hippopotami, crocodile, fish and other aquatic creatures. Obviously this area has not always been a desert. Satellites have revealed a vast drainage network of freshwater lakes and channels underneath the sand dunes. Countless tools, pottery, harpoons and thousands of pictures of animals carved on rocks have been found, showing that humans also lived in this once tropical region. It would seem that "climate change" is nothing new.

The Great Basin of Southwestern United States, which is now very dry, reveals that it was once well-watered, having some 120 lakes.



The Great Salt Lake in Utah was about the size of Lake Michigan. Ancient lake shorelines are etched high up on the sides of hills and surrounding mountains. Death Valley once possessed a lake about 600 feet deep.

After the flood when the animals left the ark, they needed a well-watered place in which to feed and reproduce. During the Ice Age (4000 years ago), areas of the world that were not cold enough for heavy snow falls received generous amounts of rain for hundreds of years. The humid, wet climate was ideal for rapid plant growth and produced much vegetation for 500+ years following the Flood. God's mercy and provision is even revealed in the pattern of the world's climate following the Great Flood catastrophe.

**And the dove came in to him in the evening; and, lo, in her mouth was an olive leaf plucked off: so Noah knew that the waters were abated from off the earth. - Genesis 8:11**

Humans can make over 50 different distinct sounds in speech. These sounds are combined to make thousands of words in a language. To make these sounds requires precise design features and combination of a throat, muscles, tongue, and a speech processing center in the brain. Let's look at these design features in greater detail:

- A deep throat as compared to an ape's shallow throat. A deep throat allows a greater variety of sound to be produced.



- Fine muscle control of the vocal tract. Humans have over 100 muscles to control the vocal chords and tract, while apes have far fewer muscles.
- Uniquely agile tongue and lips. Humans can shape many sounds because of their agile tongue and lips. Apes do not have fine control of their lips and tongue, thereby limiting their range of sounds.
- Language center in the brain. Humans have areas in their brains dedicated to processing language; apes do not.

We are very different from apes in the area of language and speech. The ability of humans to communicate is unique as compared to all animals on Earth. We were designed to verbally communicate thoughts and ideas. The skill of human speech is just what we would expect from the Creator who made us in His image to communicate. Of course, the most important use of language is to give thanks to and share truths about the awesomeness of our Creator, Savior, and Lord God – Jesus Christ.

**Surely I would speak to the Almighty, and I desire to reason with God.**

**- Job 13:3**

What weighs 10 million pounds but can float in the air? Believe it or not, it is a typical sized cloud! The total amount of water continually suspended in the air is estimated at over 50 trillion tons! What a marvel of sheer engineering! What principle of physics keeps clouds from falling out of the sky, and what is the recipe for making a cloud?

- Heat from the sun changes liquid water from lakes and oceans into water vapor – an invisible gas.
- When the sun heats the gas molecules, they rise.
- As it rises, the water vapor slowly cools.
- When the vapor touches tiny pieces of dust in the air, it turns back into a liquid.
- When billions of these droplets of water come together, they form a cloud.

As long as the cloud is warmer than the air around it, it floats – not unlike a giant hot air balloon. Gravity is always slowly pulling the cloud down, but heated air from the sun pushes it back up. Clouds are made of either liquid water or ice particles. Both are heavier than air and are constantly falling toward Earth, but most



of the time, they fall very slowly and either evaporate long before they reach Earth or are pushed up by air currents. When enough weight accumulates or air currents allow, rain or snow is the result. Our planet has a total average precipitation that would cover the entire Earth to a depth of three feet. This supply of water in the atmosphere is maintained by the power of the sun. We take this all for granted, yet God set this up in order to water the Earth. God told us of this matter thousands of years before we really understood it. We did not even know how much clouds weighed until quite recently – but God knew!

**He bindeth up the waters in his thick clouds;**
**and the cloud is not rent under them.**
**- Job 26:8**

# INSPIRED BY BIOLOGY

MAY 12TH

God has designed the chicken's egg to be a perfect nursery. Inside the egg, the nutritional yolk provides food for the developing chick. The egg white is also a food source and acts as a cushion, absorbing any bumps when the egg is turned. The egg has a strong shell for protection. In fact, architects have copied the egg shape in building domes, such as on the United States capitol building.

Imagine an egg without a shell – the egg would be squished during incubation. Yet the shell is not solid but porous – allowing the chick to breathe. A typical egg has 10,000 tiny pores. If there were too many pores, then the egg shell would be too brittle, and the egg would break too easily. If there were too few pores, the chick would not get enough air and would die in the shell. Evolutionists say that the number of pores happened by accident and chance. If this was true, we would have lots of dead, undeveloped chicks long before the bird could have ever evolved. God designed the exact number of pores for the exchange of oxygen and carbon dioxide in order for the chick to develop and hatch.

On day 19 of a chick's development, it is too large to receive enough oxygen through the pores. It is time to break out, but how, when the egg shell is so strong? Miraculously, the chick has been developing just the right escape equipment – an egg tooth. The egg tooth is a ridge on the chick's bill used to pierce through a membrane within the shell that hides a six hour supply of air. Does the chick now relax since it has received a new supply of air? If it did, the air would run out, and the chick would die. No, it uses those critical six hours to peck through the shell and gasp its first breath of outside oxygen. How does it know to do this?

The egg is the perfect nursery for the developing chick, providing food, protection and air. When we see the design of an egg and the instincts of a chick about to hatch, we know there must be a Designer.

*I will speak of the glorious honour of thy majesty, and of thy wondrous works.*
*~ Psalm 145:5*

Astronomers are finding that our sun is no ordinary star. Our sun is a yellow dwarf, and yellow dwarf stars are a minority. Red dwarfs make up 80% of stars but are not conducive to life because they emit most of their radiation in the red part of the spectrum, which make photosynthesis less efficient. For life to be possible on any planet orbiting a red dwarf star, it would need to be closer to the star. But the closer planet would tend to stop rotating – resulting in the lit side being too hot for life, and the unlit side being too cold.

Red dwarfs also have flares. For this kind of star, flares cause the star's luminosity to vary. With flares come dangerous radiation. Astronauts in orbit always take cover in shielded parts of the space station whenever this happens. On Earth, we recieve very minor



effects from this deadly particle radiation because the Earth's magnetic field funnels it to the poles. Without our strong magnetic field, this particle radiation would eventually strip away our atmosphere, resulting in the Earth becoming a dead, sterile planet.

Our sun has the right mass, the right light, the right composition, the right distance, the right orbit, the right galaxy and the right location for living beings. The vast majority of stars would be automatically ruled out for life. That makes our sun and Earth rare indeed.

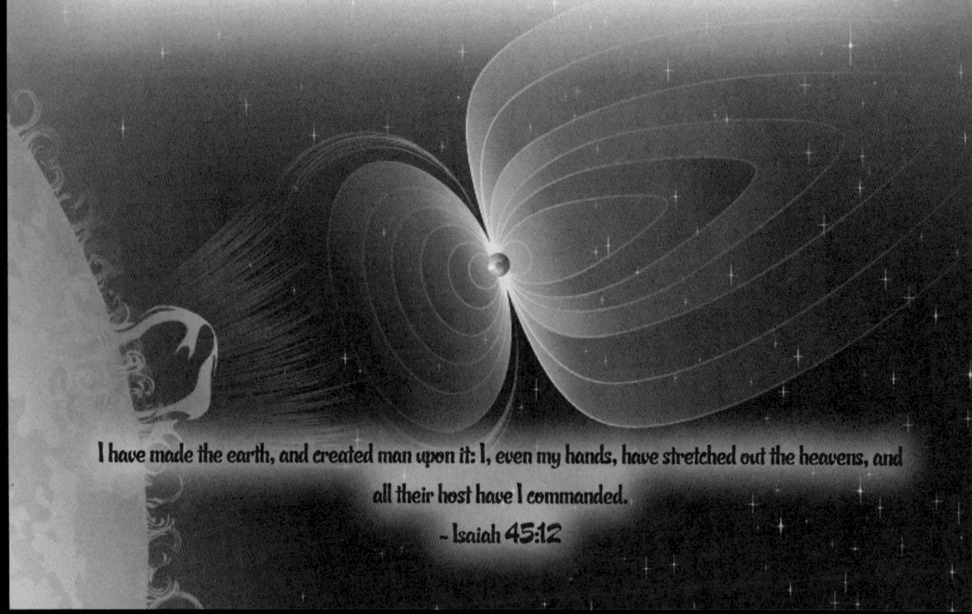

I have made the earth, and created man upon it: I, even my hands, have stretched out the heavens, and all their host have I commanded.

~ Isaiah 45:12

Those determined to explain all of biological life without God use natural selection. According to evolutionists, it is natural selection which has transformed bacteria into fish, fish into frogs, lizards into mammals, and ape-like creatures into people.

Let's look at a manufacturing analogy. Suppose I wanted to make a series of wagons, but each subsequent wagon had to use the manufacturing details used to produce the previous model. In other words, before producing the second wagon, the factory must be rebuilt using the information from the first wagon. Mutations are exactly like randomly rearranging a few letters from the factory manual for each new wagon generation. Before producing the third generation of wagons, hundreds more letters in the manual are randomly rearranged. Will this process ever allow the wagon factory to start turning out rocket ships? Obviously not! What we would see is a downward degradation of the wagon.

Now, let's switch to your body. It requires genetic information - sort of like a manual for making your hands and feet. The same is true for any creature or plant. Evolutionists will say new creatures were formed through natural selection, but nature cannot simply produce new features. Nature works with what is already present.

If you believe fish turned into frogs, the information had to come from somewhere. Random changes to the fish code (DNA) will not turn the fish into a frog any more than random changes in a



wagon factory's operating manual will allow it to start turning out rockets. So where did all the variety of living things come from? Not natural selection, but from the God of the universe.

*For we know that the whole creation groaneth and travaileth in pain together until now.*

*- Romans 8:22*

What happened to the dinosaurs? The Bible indicates that land animals were made on day six of creation week. Dinosaurs were land animals. Man was also made on day 6. Therefore, dinosaurs were made along with man (Genesis 1:24-25). Dinosaurs and man lived at the same time. When God called Noah to build an ark, representatives of all the kinds of air-breathing land animals, which included the dinosaurs, went on the Ark. Those dinosaurs left outside Noah's Ark died in the horrific flood, and many of these became fossils. After the Flood, the animals saved on the Ark (including the dinosaurs) disembarked and attempted to survive in the new environment.

As dinosaurs left the Ark, they experienced a totally different world from the pre-flood world. The climate was much harsher, and the Earth's atmospheric pressure and composition may have been much different. These changes, combined with hunting by mankind, resulted in many types of dinosaurs rapidly going extinct.

If people lived with dinosaurs, you would think that ancient historical writings, like the Bible, would mention them. And they do, just using a different word. The Bible was translated into English in 1611, but the word dinosaur was not coined until 1842 – a 231 year difference. However, the word "dragon" is repeatedly mentioned throughout the Bible (at least 21 times) and in ancient records from around the world.[1] Dragons were dinosaurs. When we view the



world through Biblical glasses, all observations – even the co-existence of dinosaurs with mankind and why dinosaurs no longer exist – make perfect sense.

...I will make a wailing like the dragons...

~ Micah 1:8

A fossil is a rock. It takes very special conditions to make a fossil. Think about this: what happens to a dead animal? Scavengers eat it, and bugs/bacteria cause it to rot and decay, eventually leaving no remains behind. It takes very special conditions in order to make a fossil. Here is the recipe for making a rock containing a fossil:

- Fast coverage by sediments, so scavengers and bacteria don't eat it.
- Deep coverage by sediment, so no oxygen is present to start decay.
- Lots of water, so the minerals can seep into the bone and turn it into stone.
- Lots of minerals in the sediment for cell replacement.

What event in history had fast, deep coverage with lots of mineral-filled water? The Flood of Noah's time. Fossils remind us of God's judgment and God's mercy: God's judgment – He destroyed an



increasingly evil and sin-filled world with a worldwide flood, and God's mercy – He saved Noah, a righteous man, and his family on the Ark.

Every time a dinosaur bone is dug up, it reminds us of God's judgment and God's mercy; fossilized dinosaur bones are really missionaries to the world.

**And the Lord said, "I will destroy man whom I have created from the face of the earth; both man, and beast, and the creeping thing, and the fowls of the air; for it repenteth me that I have made them."**

**- Genesis 6:7**

The vampire bat survives by stealing and drinking the blood of large mammals, but has this always been the case? Creation biologists have long wondered how vampire bats started to crave and drink blood when the Bible clearly teaches that there was no death or bloodshed



before the Fall. Some scientists have suggested that insect-eating bats developed this blood sucking habit accidentally while picking insects off of animals. When removing the insect, they may have accidentally wounded the host animal and, thereby, tasted some of its blood. Eventually they switched from insect-eating to exclusive blood-drinking.

Interestingly, the vampire bats are unique among bats in that they have certain features that made it easier for them to develop the practice of drinking blood. Vampire bats already have sharp teeth, which probably were used for eating fruit and would need just a little modification for piercing flesh. The tongue structure is also well-suited for lapping blood. These vampire bats live only in the tropical regions of Central and South America. Another creature that once ate insects but now drinks blood is one species of Darwin's finches. It has been observed attacking other birds and drinking their blood. However, just because we live in a fallen world where this type of survival is part of nature does not mean things were originally designed to operate in this way.

In God's original creation, everything was made "very good" even vampire bats! Death is a result of mankind's rebellion.

**And God saw every thing that he had made, and, behold, it was very good.**

**~ Genesis 1:31**

The hoatzin is a strange bird that lives in South America. This bird has blue skin, red eyes and a four inch Mohawk of spiky feathers on its head. Young hoatzins have claws on their wings, so they can climb trees like monkeys. They are expert underwater swimmers even before they can fly. But that is not the strangest thing about this bird.

Ninety-five percent of its diet consists of leaves and vegetation. Most birds survive on nuts and bugs. The hoatzin is the only bird known to digest its food like a cow. Like other ruminates, the hoatzin has a special chamber with bacteria to break down plant

## INSPIRED BY BIOLOGY

MAY 18TH

material. The question remains, how did the bird learn to do this? Is this bird's closest descendant a cow, or did a cow decide to turn into this tropical bird one day?

How do evolutionists explain this oddity? One evolutionary scientist who has studied the hoatzin for decades said, "Hoatzins don't seem to follow the rules of evolution." Hoatzins defy evolutionary explanation. When God designed the hoatzin, He designed a creature that defies any human explanation. Maybe He wanted us to know the truth – evolution can't explain the hoatzin…because evolution from one creature into a completely different type of creature (macro-evolution) has never happened.

The Lord Almighty is truly a creative, imaginative God.

**…He is a great King over all the earth.**

**- Psalm 47:2b**

Most consider Moses the author of Genesis. Jesus even used the writings of Moses to prove that he was the Messiah. But what method



did Moses use to assemble the knowledge of the first five books of the Bible? These are the several possible options – (1) direct revelation from God, (2) oral traditions, (3) written records of the past, placed into a final form as guided by the Holy Spirit.

The third possibility seems the most likely. In Genesis, we find the Hebrew word "toledoth", or generation, in 11 different places.

- "These are the generations of the heavens and of the Earth" Genesis 2:4
- "This is the book of the generations of Adam" Genesis 5:1 (the word "book" makes it clear that it was actually written down).
- "These are the generations of Noah" Genesis 6:9 (Noah compiled these records)
- "Now these are the generations of the sons of Noah" Gen. 10:1
- "These are the generations of Shem" Genesis 11:10
- "Now these are the generations of Terah" Genesis 11:27
- "Now these are the generations of Ishmael" Genesis 25:12
- "And these are the generations of Esau" Genesis 36:9
- "These are the generations of Jacob" Genesis 37:2

The book of Genesis was likely compiled from actual eyewitness reports of actual events. These accounts were possibly recorded on tablets of stone or clay and handed down from father to son, finally coming into the possession of Moses. Moses selected appropriate sections for compilation and inserted his own editorial additions which provided a smooth transition from one document to the next, with the final result being the Book of Genesis. (The Jews have a tradition that Adam wrote a long book about God, but it has been lost, and we only have this short portion). We have the privilege of reading this ancient history book.

**And beginning at Moses and all the prophets, he expounded unto them in all the scriptures the things concerning himself.**

**- Luke 24:27**

Almost all forms of complex life have both male and female. Each needs the other in order to make offspring. The male needs the female, and the female needs the male. If evolution were true, which came first? What if the female evolved many years before the male? No babies would result. What if they both evolved at the same time, but in different parts of the world? Again, no babies. How could both the male and female reproductive systems evolve at the same time, each system being complex, complimentary, and different?

## INSPIRED BY BIOLOGY

MAY 20TH

If you observe our modern culture, you will notice an attempt to try and blur the differences between men and women, as if we can choose which we want to become. Yet it never works – males were created to be males and females to be females – with vast physical, psychological, biological, emotional, and chemical differences.

Only the Bible gives a satisfactory explanation for the sexes. Jesus speaks in Matthew 19:4, "at the beginning the Creator made them male and female." From the beginning of creation, God made males and females in order to propagate both the human race and animal kinds.

**So God created man in his own image, in the image of God created he him; male and female created he them. And God blessed them, and God said unto them, "Be fruitful, and multiply, and replenish the earth..." - Genesis 1:27-28**

How long does it take a fossil to form? People are conditioned to think about millions of years and evolution whenever they hear the word "fossil." But is this really a valid conclusion or just storytelling?

Near York, England is the Dripping Well of Yorkshire. This place is famous for turning soft cuddly teddy bears into stone. Since the 1600's, this has been a tourist attraction where people have hung clothes, hats, shoes, and, yes, even teddy bears under a waterfall that has turned each into stone.

The waterfall's water originates underground and has an extremely high mineral content. As the water splashes onto the hanging objects, the mineral



calcite (calcium carbonate) is deposited along with small amounts of other minerals. Over the months, these deposits build up and coat the object with a crust of rock. Petrifaction time depends on the size and porosity of the object. Small teddy bears turn to stone in three to five months. Larger teddy bears take six to twelve months.

This is not the only place in the world where petrifaction has been observed. Australia has its own petrified waterwheel that has become totally encased in stone in only decades, while New Zealand has a petrified bowler hat on display. It does not take millions of years for petrifaction; it just takes the right conditions.

*Thou coveredst it with the deep as with a garment: the waters stood above the mountains. At thy rebuke they fled; at the voice of thy thunder they hasted away. ~Psalm 104:6,7*

Truth is a concept under siege. Like the last stand at the Alamo, our culture seems determined to destroy every last remnant of the belief in truth, and Christians seem besieged on all sides by overwhelming forces of deception. Consider the adjectives commonly



coupled with the concept of truth – 'partial' truth, 'half' truth, 'gentle' truth, 'relative' truth, 'debatable' truth, 'white' lie, 'sort-of' true, 'your' truth, 'near' truth, 'grey' area.

In actuality, these words coupled with truth are self-contradictory. If truth is relative, then truth does not exist. Society cannot function without the existence of "true" truth. Furthermore, the claim that "truth is relative" is in itself an appeal that absolute truth exists. Truth (or an equivalent term) is mentioned at least 389 times in the Bible. Just before his brutal death, when asked by Herod who he was, Jesus responded by stating that his purpose in coming to Earth was, "to testify to the truth (John 18:37)." Truth is absolutely central to understanding God, Jesus, Satan, the Bible, and the cosmic battle we are engaged in.[1]

If we cannot trust the clear straightforward statements of the Bible to be true, we truly are lost upon the shifting sands of relativism. Truth cannot contradict itself, and there cannot be alternate truths. This is because God is truth, and just as there is only one Creator, there can be only one truth.

**Jesus saith unto him, I am the way, the truth, and the life: no man cometh unto the Father, but by me.**

**- John 14:6**

Have you considered dust as a blessing from God? Yes, the dust that accumulates on your furniture; the dust that makes you sneeze.

Every drop of rain is gathered around dust. If the air is too clean, rain has a hard time forming. Dust is needed. If there were no dust, then there would be no rain, no plants, no animals, and no you!

In addition, it is the dust in the air that contributes to gorgeious sunrises and sunsets. Without dust, the beauty we see on the horizon at the beginning and ending of each day would be greatly diminished.

So the next time you are frustrated with all the dust on the furniture, stop and thank God for it – insignificant in size but very significant for life.

May 23rd

**DESIGN** INSPIRED BY

**And the LORD God formed man of the dust of the ground, and breathed into his nostrils the breath of life; and man became a living soul. ~ Genesis 2:7**

Have you considered how marine birds get their drinking water? Their only nearby source is salty ocean water. But wait you say, everyone knows that if you drink ocean water, you will die. So why don't these marine birds die?

These birds are equipped with their own desalinization plants in their noses. The salt gland filters out the extra salt from their blood. The concentrated salt water is then eliminated by a duct that runs to the nose. Often you will see a marine bird with a runny nose – this is just concentrated salt water running down the



bird's bill. Or you may see a bird sneeze; this is another way to eliminate the salt water.

Interestingly, birds such as the loon that live part of their life on fresh water and then migrate to the oceans for winter also have salt glands. When the loon returns to fresh water in the spring, the salt gland shrinks. But the moment that it drinks salt water in the fall, the salt gland swells up, and it starts processing again!

Around the world are thousands of desalination plants for people. These plants take salty sea water and process it to become fresh drinkable water. They require carefully planned technology and cost millions of dollars to build. Did these plants build themselves over time or develop by accident? We look with pride at what we have created and how we have provided people with drinkable water. God looks with loving care at how He has provided for His creatures.

**Let your conversation be always full of grace, seasoned with salt, so that you may know how to answer everyone. - Colossians 4:6**

Geologists from around the world descend upon the Grand Canyon to see the layers of rocks. Few places on earth reveal such a sequence of rock layers. But what do these layers tell us?

- Within all the horizontal layers are marine fossils such as corals and clams. Marine fossils are found on every continent in the world.
- One layer in the Grand Canyon is called the Coconino Sandstone or "the bathtub ring." This formation has been found across the Unites Sates but given different names – Colorado and Kansas: Cedar Hills Sandstone; Oklahoma: Duncan Sandstone; Texas: Glorietta Sandstone.
- The Tapeats Sandstone, one of the lowest layers of the Grand Canyon, appears to form a single sandstone body that blankets across North America, extending from California on the west



northward to Montana, covering much of North Dakota, and across the Mid-west to New England. The same formation in southern Israel is called the Nabetean Sandstone of Petra.

- The Redwall limestone in the Grand Canyon is grey limestone stained red from the iron oxide in the above layer. This same limestone can be found in much of North America, England, and in the Himalaya Mountains.

During the Genesis Flood, there was catastrophic flooding across the earth. These tidal waves and tsunamis killed marine creatures and poured onto the land. Subsequently, tsunamis laid down more layers of fossil-containing sediments. The Flood continued for months. The Grand Canyon reveals the violent worldwide flood that once was; it is a monument to the Genesis Flood.

**Noah was six hundred years old when the flood of waters was upon the earth.**

**~ Genesis 7:6**

The Grand Canyon is known for the rugged beauty displayed in its landscape. When gazing at the sedimentary rock layers of the Grand Canyon, notice the layers. How did they form? The dominant view today is that slowly and gradually sediment was laid down over millions of years. If this were true, we would expect to find weathering and erosion in between the layers, yet we do not. What is observed is near-flatness with knife-edge boundaries between the layers and very little evidence of gullies or other erosional features. Also, there is a lack of soil with each layer. If each layer represents thousand of years, where's the soil?

Consider what tree roots, gophers, worms and other animals do to alter the surface layers in short order. They significantly mix the soil in as little as 18 months time. This is called bioturbation. All their biological activity mixes up the soil resulting in lack of layers. None of this mixing is seen – distinct sedimentary layers are observed.

## INSPIRED BY GEOLOGY

MAY 26TH

A cataclysmic global flood as described in Genesis 7-8 is a far superior explanation for the layers we observe. The flood waters would have swept over the continents, catastrophically eroding sediments, transporting them and then rapidly depositing them in distinct flat layers upon the newly scoured surfaces. This would have continued for over a year. Caught in this sediment would have been billions of plants and animals rapidly buried with the sediment. The evidence of layers being nearly flat and having knife-edge boundaries declares that the Genesis Flood did happen, just as God has told us.

**For if God spared not the angels that sinned...And spared not the old world, but saved Noah, one of eight people, a preacher of righteousness, bringing in the flood on the world of the ungodly...**

**~ 2 Peter 2:4,5**

The Earth is halfway between the center and the edge of the galaxy, between two of the spiral arms. This is the perfect location. If the Earth were near the center of our galaxy, we would be in constant peril. This star-packed region is full of exploding stars, black holes, and deadly radiation. Earth's location has the minimum number of threats allowing maximum habitability.

The Earth's location is also an ideal platform for observations of the universe because it presents a clear view of both our galaxy and the rest of the cosmos. If we were in the center of the galaxy we would not see the rest of the universe because of the great number of stars and dust. If we were farther out or at a different angle, we would not have a clear picture of our own galaxy.



## COSMOLOGY INSPIRED BY

Why would we want to study the universe? To comprehend the enormous variety and beauty of all that God has made! To gain a picture of God's power, creativity, and engineering skills! Astronomers are finding out that the universe is finely tuned to perfection. Why is that? The Craftsman of the universe has made something that works well, and this Craftsman wants us to see a glimpse of His mind as we observe all that He has made. We are looking into the mind of God as we study the universe.

*He revealeth the deep and secret things: he knoweth what is in the darkness, and the light dwelleth with him. - Daniel 2:22*

The pitcher plant is a carnivorous plant that traps its prey. The leaves form a deep cavity in which a liquid is at the bottom, like a pitcher of water. To catch bugs, sweet nectar is needed. Instead of mixing the sweet nectar with the water at the bottom of the pitcher, this plant drips it down the outside of the container. An unsuspecting ant walks up the plant, slurping the sweet nectar. Then the ant arrives at the top of the pitcher, but it does not stop because the plant has nectar along the inside rim. It's almost like free candy. As it slurps up this sweet nectar under the rim, it moves farther into the plant until it arrives where the walls of the container are very slippery. Ants that can even walk upside down on a ceiling are helpless here.

It is just too slippery and so it slides down until its feet catch on stiff hairs of the plant. These hairs are pointed downward toward the pool, so as the ant struggles to hold on, he steps closer and closer to the pool. These hairs signal the plant to get ready for dinner. Digestive juices start flowing. The inside of the pitcher plant becomes slipperier as the ant struggles harder and harder. Finally the ant loses the battle and falls into the pool of death and drowns. The digestive juices, already in the water, dissolve the ant and allow nutrients to be absorbed by the plant.

This amazing mechanism is said to happen by accident and chance. How did the plant know it needed sweet nectar to attract insects and then to lure them by putting it on the outside and inner rims? How did the plant know to point the hair downward and not upward? How did it know to activate the digestive juices only when the hairs are stimulated and thereby not waste the digestive juices? How did it know how to make digestive juices – remember this is a plant and not an animal! There are far too many features present to have evolved over eons for the pitcher plant to live. All these features needed to be present and fully functioning from the moment the pitcher plant was designed. This mechanism for survival was needed only after sin entered the world. God knew what the pitcher plant needed and designed it to survive.

**The earth is the Lord's, and the fulness thereof; the world, and they that dwell therein. ~ Psalm 24:1**

A sac called the pericardium is a tough, thin, fibrous membrane which surrounds and protects your heart. Imagine it this way; your heart is in a plastic sandwich bag with another sandwich bag surrounding it, and between the two bags is an oily lubricating fluid. Your heart is essentially encased in this tough sac with the lubricant fluid between the heart and the sac. The pericardium must also beat in rhythm with the heart. Heart palpitations occur when this



is out of sync. Often the pericardium is beating at twice the rate of the heart. One cause of such palpitations are electrical problems created by a shortage of magnesium. But why is there this oily fluid surrounding the heart?

When the heart beats, it can now slide around in the fluid without creating friction. Friction, like the rubbing together of your hands, produces heat. A heart without this marvelous sac (pericardium) would soon produce enough heat to kill us. How do evolutionists explain this? During the process of evolution, how many hearts would have died from over-heating because the sac had not yet evolved? What about the fluid in-between? When would that have evolved? A heart without a fluid-filled sac is an overheated heart. Both had to have appeared at the same time. These kinds of details are ignored by the evolution believers. They just wave a magic wand and say, "Evolution did it."

God designed the lubricated sac (pericardium) to be fully functioning from the beginning in order for the heart to survive.

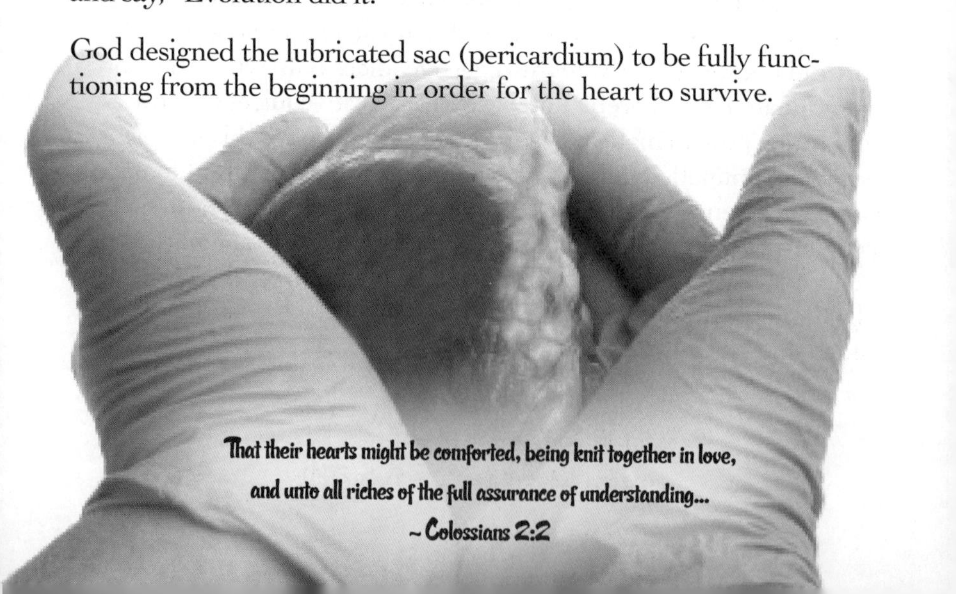

**That their hearts might be comforted, being knit together in love, and unto all riches of the full assurance of understanding...**

**~ Colossians 2:2**

The desert is an extremely harsh environment, so how do animals survive? Let's consider two of them: the jackrabbit and the kangaroo rat.

The jackrabbit has large, floppy ears. Biologists once thought that large rabbit ears were only used to enhance hearing. However, we now know that their ears perform an even greater function. Scientists investigating heat-stressed jackrabbits found that the blood leaving the ear was cooler than the blood entering the ear. They found that when the jackrabbit was too hot, it would



expand the blood vessels within its ears, allowing more blood to flow through and thus removing body heat. The large ears act like radiators! During the midday desert heat, a jackrabbit sits in the shade of a bush with its ears erect, radiating heat away from his body and keeping himself cool and comfortable.

The kangaroo rat conserves water because open sources of water do not exist within its desert habitat. As the kangaroo rat digests his food, his very cells capture and reuse the normal waste product of the food breakdown process – water. Then he conserves this water by being nocturnal, sweating little, and having extraordinarily efficient kidneys.

These desert creatures are obviously designed to survive and even thrive in such harsh conditions. Do random chance mutations or a powerful Designer seem like the better explanation for these complex survival mechanisms?

**Then shall the lame man leap...and the tongue of the dumb sing: for in the wilderness shall waters break out, and streams in the desert.**

**- Isaiah 35:6**

At the Tower of Babel, one common language of mankind was turned into many languages. Because of their inability to communicate with each other, the people were forced to spread and fill the Earth. Those who arrived in China could still remember the true history of the world: the creation, the Garden of Eden, the Flood and the Tower of Babel. For nearly 2,000 years, the ancient Chinese worshipped the true God, Shang Ti.



In the early days of this civilization, someone had to write down the new Chinese language. This unknown linguist settled upon a pictorial symbolism for their letters. For example, the inventor had to figure out how to write the word for "boat". He chose three symbols for boat: the symbol for vessel, the symbol for eight, and the symbol for mouth. Mouth stands for a person even in our language. For example, we say a family has five mouths to feed. So the symbol for boat was a vessel with eight people! The inventor of the early Chinese language was obviously thinking about the most famous boat he knew – Noah's Ark with eight people on board.

There are actually dozens of the original Chinese symbols which exactly correspond with events right out of Genesis chapters 1-11. Embedded within the Chinese language is the history and the knowledge of God and the true history of mankind up to and including the Tower of Babel.

**There are, it may be, so many kinds of voices in the world, and none of them is without signification.**

**~1 Corinthians 14:10**

中擇其精者二百幅編為榮寶齋詩箋譜也
版這是中國版畫史上的計程碑之一自有他的音
義與價值在今日也還值得再版特別是北平
箋譜絕版已久無法再印有此一書探討三四十
年前版畫史的人也就有一部份材料可以依據了
我們應該歡迎這部書的再版出來一九五一年
六月二十三日鄭振鐸序

鄭振鐸

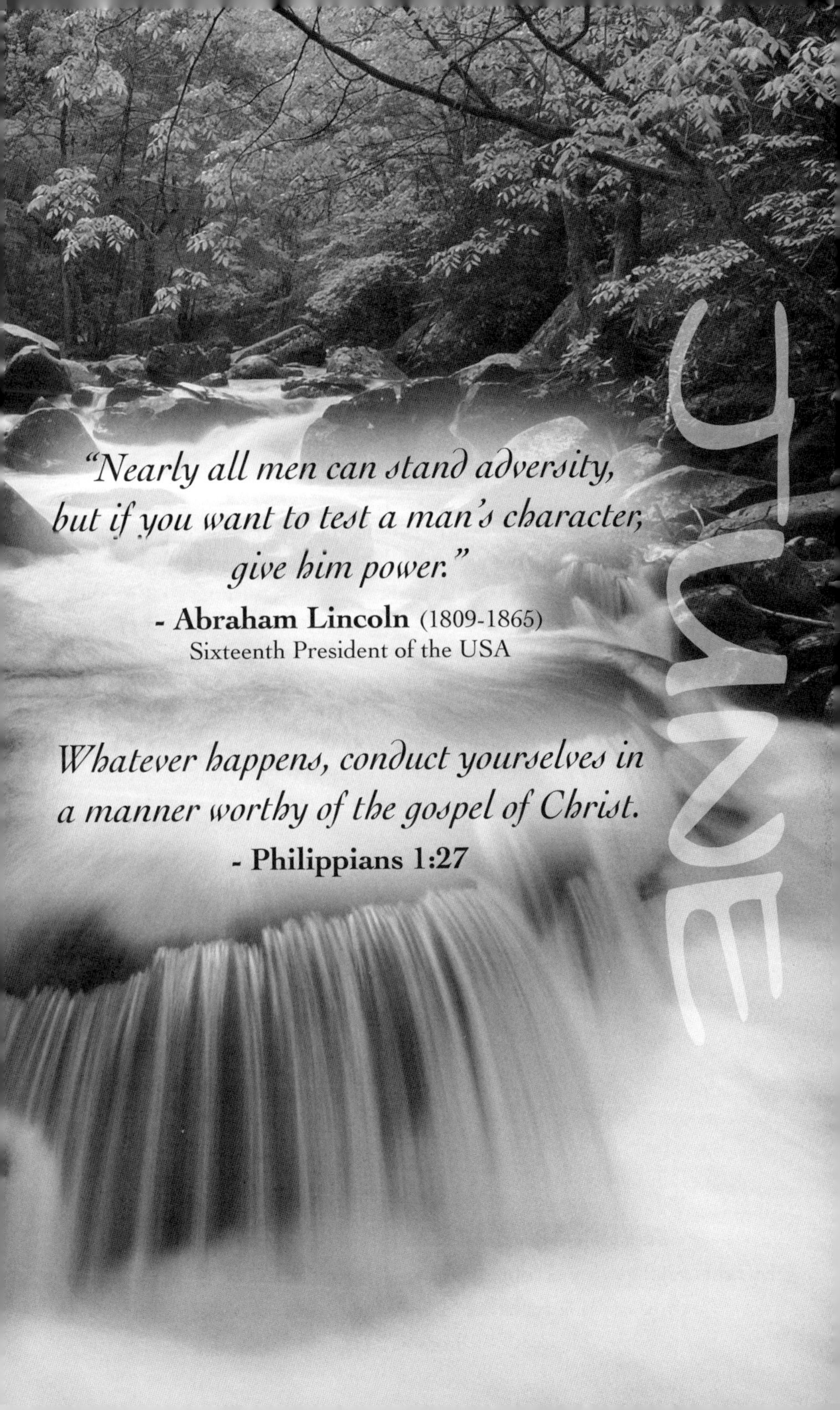

# JUNE

*"Nearly all men can stand adversity, but if you want to test a man's character, give him power."*

**- Abraham Lincoln** (1809-1865)
Sixteenth President of the USA

*Whatever happens, conduct yourselves in a manner worthy of the gospel of Christ.*

**- Philippians 1:27**

The polar bear has amazing fur! A number of years ago, researchers needed to determine how many polar bears lived in the Arctic. Scientists flew over the region using an infrared detection system to count the number of bears. Surprisingly, none were found. How could this be? They knew polar bears were down there, but why were they not detected? Nature had "out-smarted" the scientists!

Infrared detectors work by detecting heat which is radiating from the body. It was discovered that polar bears were so well insulated that they experienced almost no heat loss through their fur. Therefore, they could not be detected by the infrared instruments. Each hair on



the bear is hollow and filled with gas molecules. This slows the flow of heat as it escapes the bear's body. The Arctic region is bitterly cold at night, so the Master Designer had to design a fur coat that would truly withstand these brutal temperatures. Polar bear fur is unlike the fur of any other bear. Polar bears need this special fur in order to survive. It's like polar bears are covered with an extremely efficient insulating blanket – like being wrapped in a thick layer of flexible Styrofoam™. Styrofoam is the result of brilliant research and development efforts. Did polar bear fur just happen by chance? Just as Styrofoam was designed, so was polar bear fur.

P.S. Scientists were ultimately successful in counting the polar bears, but they had to switch to ultraviolet sensors.

**Every creature which is in heaven, and on the earth, and under the earth, and such as are in the sea, and all that are in them, heard I saying , Blessing, and honour, and glory, and power, be unto him that sitteth upon the throne, and unto the Lamb for ever and ever.**

**- Revelation 5:13**

After the Flood, Noah stepped out of the ark into a forbidding environment. Underneath Noah's feet were thousands of feet of

## INSPIRED BY HISTORY — JUNE 2ND

mud and sand in which the animals and plants from the pre-flood world were buried. The unstable Earth was struggling to reestablish new plants. Super volcanoes were belching ash and lava over vast regions while earthquakes still rocked the land. It took centuries for the Earth's climate to transition from the wettest period in history to today's climate.

Thousands of volcanoes and rapid continental movement during the Flood caused the oceans to warm up to 80°F (or higher), whereas the average ocean temperature today is 39°F. These warm oceans, next to cold continents, created violent storms. Enormous hurricanes (called hypercanes) drew water from the oceans and dumped it onto the land, filling depressions, causing lakes to overflow, and cutting deep canyons through newly laid sediment layers. These intense rains also saturated the ground, allowing groundwater to rapidly carve enormous cave formations; these caves would take hundreds of thousands of years to form at today's rate.

As the oceans cooled and precipitation declined, the world's reforested areas started to dry out and change to grasslands. Meanwhile, at high latitudes and mountain tops, the precipitation fell as snow and ice. Ice built up rapidly, and the Earth's one and only ice age developed, lowering ocean levels up to 300 feet! Once the oceans cooled to present day temperatures (many hundreds of years after the flood ended), the air circulation changed on Earth. This shift formed the deserts in a belt around the world, about 30 degrees north and south of the equator.

Minor climate changes (all too commonly promoted in an effort to gain political power) are insignificant compared to the dramatic changes experienced upon this planet in the last 4,000 years. The post-flood world was truly an ever-changing and dynamic place.

**And surely the mountains falling cometh to nought, and the rock is removed out of his place. The waters wear the stones: thou washest away the things which grow out of the dust of the earth... ~ Job 14:18-19**

What do the Eiffel Tower and the thigh bone have in common? The 1889 World's Fair was coming up, and Mr. Eiffel wanted to win the architect's contest by designing the most daring structure. The winning design would commemorate this great event. Gustave Eiffel went to an unusual source for his design idea: the human thigh bone. The thigh bone connects to the hip and extends sideways, causing the body's weight to be moved off-center. When the thigh



bone's head was examined internally, it was found to have beautifully curving lines from the head, while other bone fibers crossed over it. This bony crisscross pattern is like a diagram showing the lines of stress within the loaded structure. In other words, the thigh bone was strengthened in exactly the manner and direction in which the strength was required.

Mr. Eiffel simply copied the thigh bone structure using wrought-iron in place of bone and designed the now famous, flared tower. Even though Mr. Eiffel won the contest, the competing architects scoffed and predicted the tower would collapse under its own weight. It still stands today, a century later. But of course it would; Mr. Eiffel just copied what has been allowing us to stand all these years, our thigh bone. Maybe we should rename this famous structure the "Thigh Tower" … or better yet – "God's Tower". After all, the Eiffel Tower is really His design.

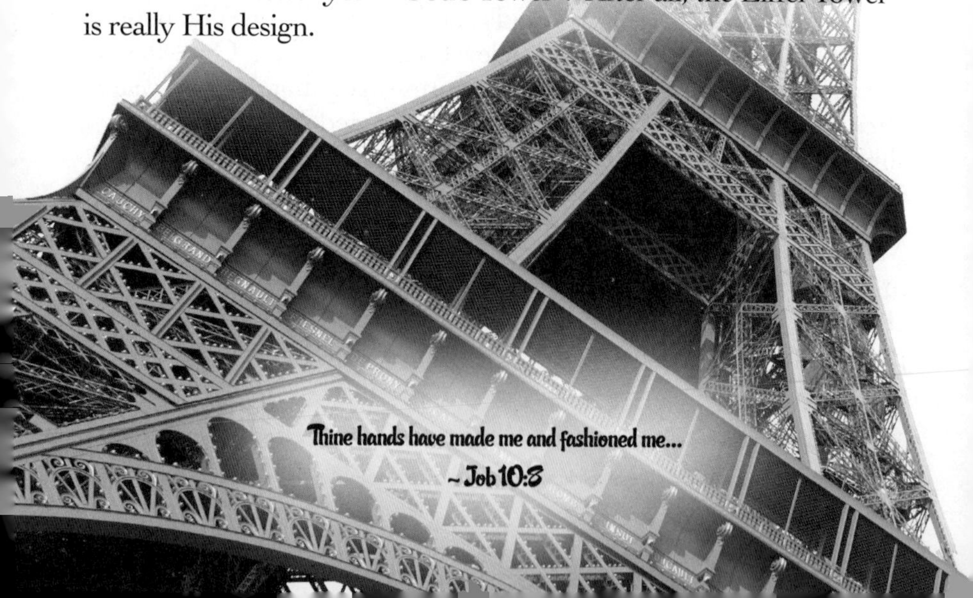

Thine hands have made me and fashioned me...

~ Job 10:8

Have you considered why fossilized human bones are so scarce? If Noah's flood really happened, would we not have found lots of fossilized human bones?

Think, for a moment, what is buried first in a tsunami type flood? It is the ones least capable of escaping – creatures found at the bottom of the ocean! That is exactly what we find; 95% of all fossils are marine invertebrates, mostly shellfish. As the flood waters rose and tsunamis rampaged, primarily ocean critters were buried and fossilized. Those that did not get covered rapidly would float and bloat, become dismembered, disintegrate, and later scavenged.

Initially, many humans would have run away from the flood and then eventually succumbed to drowning or holding onto floating vegetation until death. Most of the bodies probably would end up floating on the water, not buried in the sedimen-



tary mud. Some have estimated that at the time of the flood, there could have been 350 million people on the planet. This number of people is tiny compared to other living things. Thus, the human population was relatively small, and the chance for rapid burial and fossilization during the flood was low. Even if all 350 million people were preserved and evenly distributed throughout the world's approximate 500 million cubic miles of flood sediment, the chance of discovery, recognition, and reporting are remote. Thus, few human remains are found. The primary reason for the Genesis flood was the destruction of a widespread evil human culture, so the rarity of human fossils is no surprise.

**And God saw that the wickedness of man was great in the earth... And the Lord said, "I will destroy man whom I have created from the face of the earth." ~ Genesis 6:5,7**

From the foundation of America up into the 1960's, public school teachers were free to teach both moral and physical truths right out of the Bible. Early in the twentieth century, this was an encouraged practice throughout America, as the Bible was acknowledged as the



foundation of all reality – including science. What was the result of this practice? America became the greatest economic and technologically advanced nation on the face of the Earth. It can be argued that from 1800 to 1960, more science and medical advancements came from America than any other nation. For instance, during the first 100 years of the Nobel peace prize in science and medicine (1901 – 2001), Americans have been awarded more Nobel prizes than any other nation on Earth. The majority of Americans throughout this time period consistently stated that they believed God created humanity within the last 6000 years. This belief is clearly not an impediment to scientific advancement.

It is absurd to believe that scientific advancement will be slowed by acknowledging that the order and beauty of creation testify to the recent creation by an intelligent, all-powerful Designer. This understanding actually spurs scientific advancement by giving a firm foundation for understanding that an orderly and intelligent designer assures an orderly, understandable creation. We have the privilege to learn how this creation operates.

On the other hand, believing that the universe is just a conglomeration of random processes (such as mutations and the big bang) provides no assurance that anything has purpose or meaning. Is it any surprise that since 1961/1962 (when the Bible and prayer were removed from schools), America has fallen behind the rest of the world in Nobel prizes and science literacy…while widespread moral/social problems, such as violence, abortion, homosexuality, and personal/national debt (yes, this too is a moral problem), have seen a rapid rise?

**Blessed is the nation whose God is the Lord. - Psalm 33:12**

The heavens and the Earth declare a recent creation:

- **Too few supernova remnants.** We experience about one supernova (violently exploding star) every 25 years. The remnants of gas and dust should remain visible for millions of years. We have only directly observed approximately 200 supernova remnants. That's about 7,000 years worth of supernovas.
- **Comets disintegrate too quickly.** Comets are supposedly 5 billion years old – forming when our solar system supposedly condensed. Yet, each time a comet orbits close to the sun, it loses much of its material. Simple observation and straightforward math indicate that comets (and therefore our solar system) could not have been around much longer than 100,000 years.



- **Not enough salt in the sea.** Every year, streams and rivers dump over 450 million tons of salt into the oceans. Dividing the total amount of salt in the seas by the amount coming in at today's rate gives a maximum possible age for the oceans. (Salt accumulation may have been faster for 500 years or so following the Flood as during that time there was significantly more rainfall.) The maximum possible age for the oceans is only 62 million years.
- **DNA and other biological materials decay rapidly.** DNA experts have experimentally shown that no biologically significant lengths of DNA can exist in the natural world longer than 10,000 years. Yet, DNA has been identified in dinosaur fossils, within insects trapped inside of amber, and inside other creatures buried during Noah's flood. The very existence of this DNA is proof that these creatures were buried quite recently.

The Biblical time scale of a young world fits these and many other natural phenomena much better than the billions-of-years evolutionary timeframe.

***By the word of the Lord were the heavens made; and all the host of them by the breath of his mouth.***

***~ Psalm 33:6***

Bear Butte State Park is a designated National Natural Landmark sacred to the Native Americans. But did you know that during the 1990's the wall of this state park displayed an account of the flood of Noah?

There are some 360 flood accounts from around the world. This is the one which was recorded at Bear Butte:

"Nu-mohk-muck-a-nah (Noah) escaped a world-wide flood in a dugout canoe. [Noah] and these animals were saved and landed on Bear Butte. A dove was sent out from the canoe and it came back with willow leaves. Then, the animals left the dug-out canoe two by two."

Doesn't this sound like Noah's Flood from the Bible? How did this story show up in South Dakota centuries before missionaries brought the Bible to these people? The ancestors of the Mandan Indians brought this history with them when they spread out from the tower of Babel through Tibet into North America. Some people will say this story is from the Christian missionaries who visited the Mandan Indians. If this were true, would they not have also been told about Jesus Christ?

This type of flood rememberance would be expected from cultures around the world if an awful world catastrophe, as described in the Bible, actually happened. The story would become increasingly distorted with time, but the essence of the true, actual, historical event would remain. As you assend Bear Butte, some 1400 feet above the prairie, you can see why the Mandan Indians choose this sight for "Noah's" landing.

***We will not hide them from their children, shewing to the generation to come the praises of the Lord, and his strength, and his wonderful works that he hath done. - Psalm 78:4***

When God, the intelligent creator, made creatures He grouped them into "kinds" for the purpose "to fill the Earth". The information for reproducing after its "kind" is encoded throughout the entire organism. Each type of organism has been designed or programmed to reproduce, adapt to, and fill all potential environmental niches. This forms the basis for a more accurate term than the commonly used phrase, *"natural selection"*. Creatures actually exhibit *"programmed filling"*. This term describes how a created "kind" can fill many potential environments. With many variables within the organism, some organisms die soon while others live



a long time and have greater opportunity to pass their complete preprogrammed information on to future generations.

*Programmed filling*, rather than the evolutionary idea of n*atural selection* is the basis upon which organisms exhibit the traits necessary to live in varying environments. *Natural selection* is a misnomer. It really does not exist. The "power" for an organism to adapt is within the organism, not the environment. *Natural selection* is NOT an "intelligent force". It is NOT a creative thinker. It is NOT a talented, inanimate substitute god – as the evolutionary theory requires it to be. The design to fill different environmental niches is inherently pre-programmed within all living organisms.

**For the earth shall be filled with the knowledge of the glory of the Lord.**

**~ Habakkuk 2:14**

Have you considered how insects survive the winter? Some migrate south (like the monarch butterfly), many die, but what happens to the rest? Insects are cold-blooded, and therefore, take on the surrounding winter temperature whether it is 20°F or minus 20°F. How do insects survive these freezing temperatures to plague us again in the spring?

During the fall, insects start making glycerol (a type of antifreeze) in their bodies, and by the time winter takes place, the insect is ready. Now, as an insect's body drops below freezing, the cells do not become damaged. In fact, wooly-bear caterpillars that live in the Arctic Circle can withstand minus 90°F.

The belief in evolution requires that insects developed the ability to make antifreeze over millions of years. This is a problem because insects without the antifreeze-making ability result in dead insects. The process for making it and making the correct amount, would have been needed right from the beginning in order to survive the first winters. God made insects with the ability to produce antifreeze.

But here is the other question: Why do we need insects to survive the winter? One answer is birds. When the



birds arrive in the spring, the insects are there as food. After a grueling migration or long winter, finding insects is like a wonderful banquet for birds. God in His wisdom created insects with the ability to survive winter, and then, in the spring to proliferate in order to provide a food source for the birds.

**By them shall the fowls of the heaven have their habitation...These wait all upon thee; that thou mayest give them their meat in due season.**

**~Psalm 104:12, 27**

Fluids within all cells normally freeze, forming sharp, pointed crystals that expand and pierce the cells' membranes. This causes the fluids to leak out, resulting in death. However, Antarctic spiders, beetles, and woolly-bear caterpillars make glycerol, which slows crystal formation even when temperatures reach minus 90°F. Scientists are discovering that not only these polar-dwelling insects, but many more varieties have been designed to make glycerol or alcohol, which is similar to the antifreeze in your car. With this "antifreeze", insects can survive the winter as eggs, caterpillars, pupa, or adults. When spring comes, these insects "thaw out" and continue on in their lifecycle.

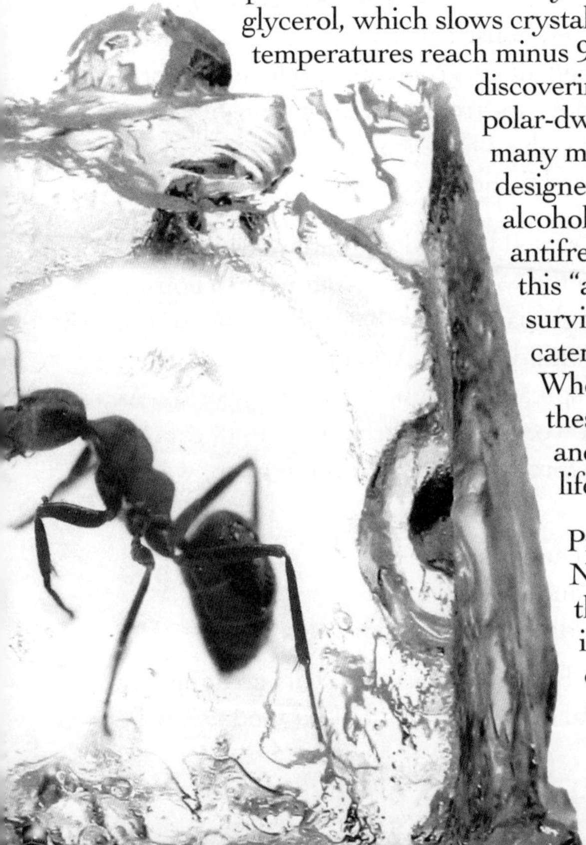

Prior to the Flood of Noah's day, it is possible that winters of snow, ice, and cold did not occur. As Noah and the animals left the Ark, we find the first mention of cold in Genesis 8. God knew

## INSPIRED BY MICROBIOLOGY

June 10th

what insects would need in order to survive, so He built in special provisions for the insects—the ability to make antifreeze and survive freezing temperatures. Jesus has also provided a way for us to truly live – through the repentance of our sins and acceptance of His sacrifice, we can "thaw out" and live in the presence of God's warming love.

***While the earth remaineth, seedtime and harvest, and cold and heat, and summer and winter, and day and night shall not cease. - Genesis 8:22***

People have always been fascinated with cats - from Siberian tigers to the common house cat. Tigers have 8 subspecies, of which three are extinct and five are critically endangered. The Siberian tiger is the largest of all wild cats - growing 10 feet in length and weighing 600 pounds. Yet, tigers and house cats have much in common. Both are independent, meticulous groomers and love to hunt, pounce and play. Tigers love to stretch out in the sunshine – just like house cats. So what is the difference between a big cat and a house cat? Big cats roar while house cats purr!

All cats may have been created by God as a biblical "kind". It is very likely that all cats in the world today, from tiger to housecat, are descended from a single pair of cats that were on board the Ark with Noah. After the cat kind left the Ark, the amazing variety manifested itself as they spread throughout the new world. We can see this biblical "kind" at work with different species of cats being interbred. For instance, through artificial insemination, we have produced:

- The liger – a lion dad and a tiger mom
- The tigon- tiger dad and lion mom
- The pumard – puma dad and leopard mom.

Even domestic cats have been bred with wild cats creating new breeds such as the Bengal or the Savannah. Did God create lions to



be meat killing machines? The original cat kind was vegetarian – as were all creatures (Genesis 1:29). Retractable claws are not just for eating meat but may have helped in extracting seeds or marking territory. Sharp teeth and powerful jaws also allowed them to eat certain kinds of tough plants. Purring cats and roaring tigers all descended from the original cat kind that God created on Day 6 of creation week.

**For every kind of beasts... hath been tamed by mankind.**

**- James 3:7-8**

# INSPIRED BY BOTANY

June 12th

One interesting way to appreciate God's design in nature is to try to solve certain problems that He had to face. Try this one. Plants need seeds to produce the next generation. But what about the insects that feed on seeds? If you decide that insects will not eat seeds, what will insects eat? If the insects eat all the seeds, how will the plant survive? So we need insects, and they need something to eat. Seeds are a convenient food for them. We have come full circle – we need seeds, and we need insects. Here are two solutions to the problem.

A plant could be designed to produce so many seeds that insects would not be able to eat all of them. Or plants could be designed to produce seeds with a natural insect repellant. Both design solutions are found in nature. Those plants that produce so many seeds that the insects could not possibly eat them all generally do not make seeds that are poisonous to insects. Those plants that produce fewer seeds often have a natural insect repellant within them. As a result, insects and plants continue on generation after generation.

Evolutionists contribute all this design to random chance changes over time - choosing to believe that because plants have survived, this proves they have evolved this ability to survive. This faulty, circular reasoning explains nothing. God in His wisdom had to solve many problems as He created the world. God displays His foresight and wisdom through the things He made.

*The works of the Lord are great, sought out of all them that have pleasure therein.*

*~ Psalm 111:2*

Michael Faraday (1791-1867) is second only to Isaac Newton as the greatest physicist who ever walked the Earth. Faraday was credited with the invention of electromagnetic induction, the electric motor, the electric transformer, the electric generator and made major contributions to our understanding of magnetism, polarized light, the liquefaction of gases, the development of rubber, optical glass, alloys of steel, electroplating, and artificial rubies. In addition, his greatest contribution to science was the development of field theory in physics. He is ranked by science historians as the greatest of all experimental physicists – adding a whole new vocabulary to modern science – anode, cathode, ion, electricity, electrode, anion, cation, magnetic field, lines of force, and electrolysis.

Faraday's work so changed modern science that two basic units of physics were named in his honor – the faraday (a unit of electrical quantity) and the farad (a unit of capacitance). Yet, as much as Faraday contributed to scientific advancement – he drew more from his deeply held Christian faith. His actions were strongly guided by Biblical truths, and his Bible contained nearly 3000 meticulously written notations in the margins. His good friend John Tyndall wrote of Faraday, "I think that a good deal of Faraday's week-long strength and persistency might be due to his Sunday exercises.



He drinks from a fount on Sunday that refreshes his soul for the week."

Just like Newton before him, Faraday drew strength and meaning from the reality that the universe displays order and meaning as a direct result of being created by God. It is not a meaningless assembly of atoms which created itself, but an orderly arrangement – designed by an incredible intelligence outside of the physical universe. This acknowledgement did not hinder the incredible discoveries of these great scientists, but provided the foundation which motivated them.

**And whatsoever ye do, do it heartily, as to the Lord, and not unto men**

**- Colossians 3:23**

One of the greatest mysteries of science is the exact structure of creation. What keeps electrons moving? What is light? What is space? What is a charged particle? Why is there so much energy contained within empty space at absolute zero (zero point energy)? Why are galaxies commonly found in a spiral shape? Why are red shifts in distinct bands (quantitized) rather than a continuous smear of shifted values? Why does the speed of light, the mass of atomic particles, and Plank's constant all seem to have a changing value with time?

## INSPIRED BY COSMOLOGY

June 14th

These are profound mysteries, and even the observation of changing values for atomic constants are frequently denied or ignored because of the implication upon traditionally accepted physics.

The primary reason modern physics is helpless to adequately explain these type of inquiries is because it starts from the wrong assumption. Modern physics accepts an enormous age for the universe and attempts to explain everything based on this starting assumption. Thus, it needs a very weak force (gravity) and lots of time to explain all of reality. Yet, this explanation cannot even explain the formation of a single star and relies on unseen faith in things like "dark matter" and "multiple universes" in an attempt to explain the impossible.

A brilliant physicist from Australia, Barry Setterfield, has developed a plasma model of physics that seems to unify the anomalies that are not explained by the Big Bang model. Setterfield's theory accepts the scriptural teaching that God stretched out the heavens and realized that this must have been accompanied by enormously accelerated radioactive decay, red shifting of light in quantized bands, and energy added to the fabric of space (analogous to how a stretched rubber band is filled with energy). According to Setterfield, this process "wound up" the universe, providing the energy necessary for star and galaxy formation – all within a very brief period of time.[1]

**Who stretchest out the heavens like a curtain.**

**~Psalm 104:2**

It is the nature of mankind to lie. The problem with lying to ourselves is that we actually know we are lying, so we must pile layer after layer of more absurd lies in a futile attempt to suppress the original lie.

June 15th

## CHRISTIAN TRUTH INSPIRED BY

Debating moral topics on an intellectual basis often drives those blind to the truth into deeper levels of denial. This is because people already know the truth. Professor J. Budziszewski teaches that people are simply choosing to ignore reality by pretending to believe something which they know, deep down, cannot be true.[1] Abortion, homosexuality, and evolution all fall into this category.

- Abortion is clearly the killing of innocent human life. A growing baby inside a mother's womb is not a blood clot, a mass of tissue, or a "fetus". It is a tiny human being. A "mass of tissue" does not do sommersaults, or suck its thumb, or have a heart that beats at 18 days.
- In a similar fashion, homosexuality is a denial of the obvious. Human bodies are designed to fit together for reproductive purposes – male into female – like a key into a lock.
- Evolution is fraught with scientific problems at every level. Matter cannot create itself, an eternal universe would be heat-dead, information (such as is found in DNA) does not organize itself by any known natural process, and most dating methods indicate a recent creation.

Only God's Word provides an unchanging plumb line to evaluate the layers of lies we hide behind. This is what is so dangerous about 'interpreting" God's Word to fit man's theories, observations, and desires. It is like bending a level to fit a crooked corner and then attempting to use the new shape to construct the rest of the house. At that point, the tool needed to evaluate the truth has been twisted and is no longer useful.

**The heart is deceitful above all things, and desperately wicked: who can know it? - Jeremiah 17:9**

The Earth is perfectly designed for life.

- The Earth is the perfect distance from the sun. The Earth is full of water. If we were too close to the sun, the oceans would dry up, the atmosphere would be destroyed and harsh rays from the sun would burn any living creature. If we were farther away, the water would freeze, and we would have icy, frigid weather all year long. There would be no growing season for vegetation. Earth is the perfect distance from the sun.
- Earth has the perfect mass. Mass determines the amount of gravity. If the Earth had less mass, it would have less gravity. When the winds would blow, things like cars and people would be blown off the ground. If the earth had more mass, an increased gravitational pull would hold harmful gases, such as methane and ammonia, close to the Earth instead of releasing them into space. Earth has the perfect mass.
- Earth has the perfect rotation. We have a 24 hour day, with about 12 hours of daylight and 12 hours of darkness. If the Earth rotated more quickly, the winds would be stronger, and there would be tremendous hurricanes everywhere. If the Earth rotated too slowly, the nights would be longer and get much colder. The longer daylight hours would cause much of the earth to have a hotter desert-like climate. The 24 hour rotational period keeps Earth's temperatures in balance to protect the plants, animals, and people. Earth has the perfect rotation rate for life.



The perfect design of Earth implies there must have been a designer, and that designer is God.

**God himself that formed the earth and made it; he hath established it, he created it not in vain, he formed it to be inhabited. ~ Isaiah 45:18**

The volcanic eruption of Mt. St. Helens in 1980 devastated 230 square miles of area and killed most of the animals and vegetation. Much of the land was sterilized by hot pyroclastic flows. It was believed that this area would take decades, if not centuries, to recover. But within a year plants began to grow, and animals returned. One such animal was the elk. After the eruption, elk were seen eating the new plant growth. These elk started having twins; their birthrate was the highest ever recorded. God designed it that way. The devastated area needed repopulation.

In the same way, after the Flood of Noah's day, the entire world had been destroyed; only those land creatures on the Ark survived. The entire Earth would have been very similar to the area around Mt. St. Helens. In Genesis 8:17, we read of God telling Noah and the animals to leave the Ark. It was God's will that these animals would spread out around the world and refill the Earth. Yet, Noah waited



7 months after landing before releasing the animals – this is just one of the indications that the flood of Noah was not a little local event – the plants needed that long to grow so that the animals could survive.

What we observe from natural disasters such as Mt. St. Helens is how rapidly such a recovery process occurs. Today, Mt. St. Helens is not a monument to disaster, but it is a monument to God's design of Earth's rapid recovery.

Bring forth with thee every living thing that is with thee, of all flesh, both of fowl, and of cattle, and of every creeping thing that creepeth upon the earth; that they may breed abundantly in the earth, and be fruitful, and multiply upon the earth.
- Genesis 8:17

Think about the many functions of our moon:

- **A "night light":** In northern countries, the full moon closest to September 22 is called the harvest moon. This occurs when many crops are being harvested. It gives farmers lots of extra light to bring in their crops.
- **A Calendar:** We get our name "month" from moon. During the month, our moon goes through its phases to document the change of days.
- **Controls our seasons:** The moon stabilizes the tilt of the Earth. Without it the Earth would swing from 0-90 degrees, and our seasons would be unpredictable.



- **Protective shield:** The moon prevents many collisions of space rocks into Earth; instead, the rocks are drawn into the moon. The far side of the moon is heavily cratered from these incoming objects.
- **It keeps the oceans healthy:** Each day tides are created by the gravitational pull of the moon. From 30% to 50% of our oxygen is provided by sea plankton. If there were no tides, then there would be greatly reduced marine plants due to ocean water becoming stagnant and overheated – reducing the oxygen supply upon Earth. We must have at least 18% oxygen in the air to survive. The level is currently a perfect 21%. In a way, the moon helps provide our oxygen.

Our moon was designed for a purpose. The name of that designer is Jesus.

*... that I may plant the heavens, and lay the foundations of the earth...*

*~ Isaiah 51:16*

Have you considered the pine cone? Conifers produce pine cones. They produce two types of cones on the same tree: the pollen cone and the seed cone. The pollen cones are usually small and inconspicuous, and once they have shed their pollen, they fall off. On a spring day when it is windy, you can see pollen clouds being blown from a pine tree. The seed cones are the larger woody cones you may associate with pine trees. Conifers depend on the wind for pollination and each species is designed to capture its own kind of pollen. According to one noted botanist, "The geometry of the cone functions like a wind turbine, channeling pollen around it in a way that maximizes pollination – so that if pollen fails to land on one scale, it will receive pollen as it is passed on from the previous."

Once the pollen reaches the egg, it takes more than a year for the egg to be completely fertilized. After twelve months of growth, the seed is finally mature. When the pinecone is ready to release its seeds, it waits for a dry, warm day. Then the pinecone scales open and the seeds (with little "wings" attached) float to a new plot of land in order to grow. If the day is cold and wet, the pinecone stays shut, waiting until the day is warm and dry and a wind can blow the seeds away. Isn't it amazing how God used the fact that wood fibers shrink when they dry to open and close the pinecone, so the ripened seeds will be released to float away on the perfect day? Pine cones are the fruit of a pine tree. As the verse in Hosea indicates, we receive our fruitfulness only from the Lord. So how are you doing as a pine cone?

In the 1400's, there were three groups of people:

1. Those who believed the Earth was flat.
2. Those who believed the Earth was round.
3. Those who did not know what to believe.



The second group acted on the truth, changed human history, and became men of renown (Columbus, Magellan & others) whose names are still remembered to this day. The other two groups may seem to have held different beliefs, but on a practical level, they both acted the same. The group who believed the Earth was flat had no interest in exploring the world because they feared they would sail off the edge. The last group was unwilling to commit to either possibility; therefore, they were frozen in place and made no significant contribution to mankind in this area. The same three choices face us on every issue involving truth.

Those who know the truth are willing to risk mockery, rejection, popularity, finances, and even their lives to defend and act upon that truth. Those who deny the truth will always oppose the truth. The third group (those who don't know what is true) simply fail to act. Both the "truth-doubters" and the "I-don't–knowers" will ultimately be relegated to anonymity because they have no motivation to act. Truth will ultimately triumph because it cannot be changed – otherwise it would not be "the truth". The truth is not subject to popular vote, majority opinion, or a democratic process. The truth simply is.

In our age it has become a virtue to pretend that truth does not exist. It is considered open-minded and 'tolerant' to accept all viewpoints rather than to search out which side of an issue is the truth, and then act on that truth. God calls us to seek out and proclaim the truth.

**I know thy works, that thou art neither cold nor hot: I would thou wert cold or hot. So then because thou art lukewarm, and neither cold nor hot, I will spue thee out of my mouth.**

**- Revelation 3:15-16**

Have you considered the fossil graveyards around the world? Countless billions of animal and plant fossils are found in extensive graveyards. Here are a few examples:

- Billions of fossilized squid-like marine creatures called nautiloids stretch for 180 miles across northern Arizona and into Nevada. Most are pointed in the same flow direction.
- Hundreds of thousands of marine creatures are buried with amphibians, spiders, scorpions, and reptiles in a fossil graveyard at Montceau-les-Mines, France.
- More than 100,000 fossils representing 400 species have been recovered from a shale layer in the Mazon Creek area near Chicago. These include ferns, insects, scorpions, jellyfish, mollusks and fish.
- The Green River Formation of Wyoming contains alligators, fish, birds, turtles, mammals, insects and palm leaves.

All over the world we find vast deposits filled with billions of dead plants and animals that were buried rapidly and fossilized in sand,

mud, and lime. Many show exquisite detail such as fish with fins and eye sockets intact. One fossil shows a fish "caught in the act" of eating another fish. A female ichthyosaur, a marine reptile, was fossilized at the moment of giving birth to her baby. Many times sea creatures are buried right along with land animals.

What evidence would you expect for a worldwide flood? There would be billions of dead things buried in rock layers laid down by water all over the world. And that is exactly what we find within the hundreds of fossil graveyards all over the world.

**They did eat, they drank, they married wives, they were given in marriage, until the day that Noah entered into the ark, and the flood came, and destroyed them all.**

**~Luke 17:27**

How do evolutionists explain metamorphosis?

**They don't even try.**

Metamorphosis begins with the mother butterfly laying an egg. The egg soon hatches into a hungry caterpillar. The caterpillar eats, and eats, and eats until it becomes a fat, wingless, worm-like creature. Then it does the craziest thing imaginable. The critter seals itself into a chrysalis and turns itself into "caterpillar soup".

Within the chrysalis everything liquefies – the caterpillar's legs, its



chewing mouth, its skin, its eyes…everything. Within this liquid, a few cells start to move. Soon some of the cells form into wings, others turn into legs, or antennas, or a sucking mouth, or eyes and so on. Eventually, a completely new creature emerges from the chrysalis – a butterfly.

How does "caterpillar soup" know how to turn into a delicate butterfly? Evolutionary scientists don't have a clue. All they can say is that the instructions for the new creatures are programmed within the caterpillars. This explains nothing. When we see a program, we know there must be a programmer. This programmer is God the Creator. He is the one who programmed all this information into the "caterpillar soup". To believe that a butterfly's lifecycle happened by "chance mutations" is remarkable. To believe that this process independently developed hundreds of times (there are easily that many different types of insects which undergo metamorphosis) is beyond credibility.

The fact that Jesus programmed the crawling, chewing caterpillar to become soup and change into a flying, proboscis-sucking butterfly is by far the best scientific answer to the origin of this creature. This rebirth is also a picture of what awaits us in heaven.

**Among the gods there is none like unto thee, O Lord; neither are there any works like unto thy works.**

**- Psalm 86:8**

Man is distinct and separate from the animals. God spoke the animals into existence but he formed man from dust. If we were to make a human body we would need, as Dr. Mayo of Mayo Clinic humorously put it:

"enough potassium for one shot from a toy cannon,
enough fat for seven bars of soap,
enough iron to make a medium-sized nail,
enough sulphur to delouse a dog,
enough lime to whitewash a chicken coop,
enough magnesium for one dose of medicine,
enough phosphorous for a few boxes of matches.
The total purchase would fill not more than a couple of grocery bags".

So there you have it: our do-it-yourself kit for making the human body – a little dust and a lot of water. But here is the problem; our kit does not have the instructions. The human body is far more complex than science can comprehend! A mere postage-stamp sized piece of skin has 3 million cells, one yard of blood vessels, 4 yards of nerves, 100 sweat glands, 15 oil glands and 25 nerve endings! And we are supposed to sit idly by while our school systems teach our children that all this happened by chance mutations and time? Hardly! God, with omniscient genius, took dust, with its odds and ends and miscellaneous chemicals, and formed mankind. The

JUNE 23RD

## MICROBIOLOGY INSPIRED BY

human body eloquently testifies to God's wisdom and power. Evolutionists prey upon our gullibility. The Bible leads us to worship.

**And the Lord God formed man of the dust of the ground, and breathed into his nostrils the breath of life; and man became a living soul.**

**– Genesis 2:7**

Have you considered your eyelids? That's what engineer Robert Kearns from Detroit, Michigan (the center of America's auto industry) was thinking about as he was driving in a misty rain. It was early in the 1960's, and Kearns was irritated that his windshield wipers had only 2 speeds - fast or slow. When he had the wipers on slow, they scraped and chattered across his windshield. When he turned them off, the misty rain was too heavy to see clearly. What to do? So he asked himself, "Why not copy the blinking of my eyes; aren't they basically wipers that operate intermittently to keep the my clear?"

When our eyes blink, saline fluid washes across our eyes to remove dust while moisturizing and protecting. Our blinking is intermittent, depending on a number of factors such as humidity and the amount of dust. Kearns began tinkering with wiper motors until they could work at intervals. This was difficult because compact integrated circuits had not yet been invented. Finally, in 1964, he patented the first intermittent windshield wiper with an adjustable delay. Today, all cars testify to his ingenuity.

The intermittent wiper did not happen by accident and chance. We know the inventor's name: Robert Kearns. When we see a windshield wiper, we know there is a "windshield-wiper-maker". In the same way when we see a blinking eye (which



is enormously more complex than Kearns's invention), we know there must be a "blinking-eye-maker", and his name is Jesus.

**For the LORD giveth wisdom: out of his mouth cometh knowledge and understanding. - Proverbs 2:6**

Samuel Findley Breeze Morse (1791-1872) was a versatile American genius who changed the world with the technology



he developed. Dubbed by many biographers as the "American Leonardo de Vinci", Morse was an artist, inventor, college professor, and writer. But first and foremost, he was a sincere Christian who steadfastly refused to compromise on his belief that the Bible was the totally accurate Word of God.

Morse is best known for his invention of long distance communication via transmission of Morse code over telegraph wires. This technology rapidly spread across the developing United States and, soon afterward, around the world as the transatlantic cable was laid upon the ocean floor. He received awards for his contribution to humanity from the kings of Austria, Denmark, France, Italy, Portugal, Spain, and Turkey. Yet, it was God whom Morse credited for His creativity. The first message ever broadcast at the speed of light through a telegraph wire was a Bible verse from Numbers 23:23.[1]

Those who promote the idea that science is incompatible with belief in God's Word display an ignorance of both history and a lack of understanding that the very founders of modern science were brilliant men with a deep faith in God's Word. This faith inspired, rather than slowed, their development of scientific principles. They knew to search out and discover God's designs and purposes.

**What hath God wrought! ~ Numbers 23:23**

Consider just a few of the ways a squirrel uses its tail:

- How does a squirrel stay balanced on a tree branch? He uses his tail like a tight rope walker uses a pole. As he moves along the branch, he tilts his tail one way or the other as the center of gravity changes to protect himself from falling.
- When a squirrel is eating, he curls the tail over his head. If a hawk decides to have him for lunch, his tail protects him as the hawk grasps a claw full of fur.

## INSPIRED BY BIOLOGY

June 26th

- As he sleeps, he wraps it around his body. In other words, he carries his own portable sleeping bag to protect himself from the cold.

Did this tail happen by chance or did the Creator design it that way? How did the squirrel know to use its tail in these ways? How did it pass this knowledge on to the next generation of squirrels?

Without his tail, the squirrel would lack balance, lack protection from flying predators, and lack the ability to withstand cold temperatures. So the next time you see a squirrel, stop and consider the marvels of his tail.

**Withhold not thou thy tender mercies from me, O Lord: let thy loving kindness and thy truth continually preserve me. ~ Psalm 40:11**

The human hand is one of the most wonderful and precise mechanical devices in all of creation. The human hand, with its amazing skill, can be seen in a wide range of uses such as piano playing, delicate touch, artwork, and construction. The human hand has 27 bones, 35 muscles and a very large number of ligaments to hold the bones together. Our hands differ greatly from an ape's hand in four areas: fine motor control, full range of motion, opposable thumb, and brain area. Consider these differences in more detail:

- Our hands have fine motor control, which means we have more nerves controlling fewer muscles. For example, in our thigh muscle we typically have one nerve controlling over 100 muscle fibers, while in our hands, we typically have one nerve controlling 10 muscles fibers. This results in fine motor control. Apes do not have this sophisticated muscle and nervous system.
- Humans have a full range of movement in their hands, from straight fingers to tightly curled fingers. Apes do not; they have naturally curved fingers and a limited range of finger movement. Apes' curved fingers are ideal for climbing in trees.
- Humans have an opposable thumb which means that each finger can touch the thumb. Try it. Apes cannot do this. Humans are able to do this because our thumb is short, and the palm of our hand is very flexible. The human thumb and finger are sized and shaped to give a perfect circular pinch grip. With an opposable thumb and fine motor control, delicate tasks can be carried out such as writing, surgery, and drawing.
- Humans also have a larger area dedicated in their brains for controlling the muscles in their hands. Apes do not have such a large section dedicated to hand control. The purpose of the human hand is completely different from the purpose of the ape hand. Human hands were designed to perform skillful tasks, while the ape's hands were designed for climbing and ambling.

There is a vast difference between the skill of human hands and ape hands. As Sir Isaac Newton said, "In the absence of any other proof, the thumb alone would convince me of God's existence".

**Thine hands have made me and fashioned me...**

**~ Job 10:8a**

As the Earth circles the sun, its north-south axis maintains a 23.5 degree tilt. It is the tilt of the Earth that controls our seasons. The majority of people asked this question will answer that the sun is closer during the summer, but this is incorrect.

The Earth is 4 million miles farther from the sun during the summer season. It is the tilt that gives us our seasons, not the distance from the sun. During the summer, the northern hemisphere tilts toward the sun, and we are warm. During the winter, the northern hemisphere tilts away from the sun, and we are cooler. Of course, those living "down under" find that the Earth is closer to the sun during the summer and farther away during the winter. This would cause the southern seasons to be much more severe than those in the northern hemisphere, but God even solved that problem.

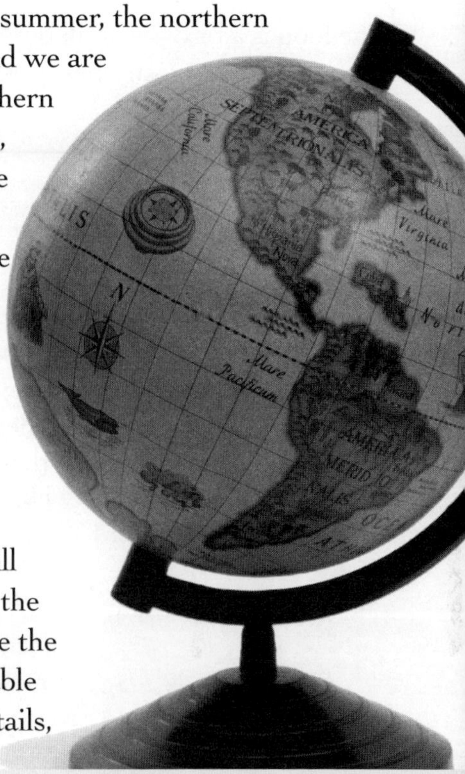


Look closely at a globe, and you will see significantly more water below the equator. This water helps moderate the seasonal changes and gives a tolerable climate year round. Even small details,



such as the tilt of Earth's axis at 23.5 degrees, did not happen by accident – Jesus designed it for our benefit.

**These are the generations of the heavens and of the earth when they were created, in the day that the Lord God made the earth and the heavens,**

**~ Genesis 2:4**

Did you know that birds have hollow bones so that they have a lower weight and fly more easily? Doesn't that sound like a logical design? Yet, how did the solid bones of a mammal or reptile, needed for strength in land animals, change into the hollow bones of birds? Evolution believers really do not know. Yet, evolutionism teaches an even more amazing story. After the solid bones became hollow – they became solid again!

The loon is a bird—and all birds have hollow bones – right? Wrong. Auklets, ostriches, penguins, and loons all have solid

bones. Consider the loon.

Loons are one of the few birds that have solid bones. Why? Because they eat fish—and in order to get these fish, they need to dive deep. Having hollow bones would be like wearing a life jacket. Imagine yourself trying to dive with a life jacket on! This would be almost impossible. God specially designed the loon with solid bones so that it could catch fish without tremendous difficulty. When you see a loon on the lake, notice how low it rides in the water—it has heavy, solid bones. In an effort to leave God out, evolutionists must believe that bird bones first went from solid to hollow and then somehow became solid again (in the loon). The next time you see a loon diving for food or hear a loon call, think of the Master Designer and His special design of the loon's bones.

**...(ask) and the fowls of the air, and they shall tell thee...**

**- Job 12:7**

Have you considered how Noah and his family cared for the animals on the Ark? The Ark was the size of a huge cargo ship (about 450 feet long). At most, 16,000 animals (2-7 of every species of mammal, bird, reptile, and amphibian on the planet – a generous approach is applied to counting the animals) would need to have been housed on its three floors. Most of the creatures would have been small – averaging perhaps the size of a small dog. Even dinosaurs would have been taken on board, not as large mature adults but as smaller juveniles. Creation scientists have calculated there are, at most, fifty different "kinds" of dinosaurs; therefore, only 100 relatively small, young dinosaurs would have been on board.

The Ark was a temporary emergency home for one year. Food and water was probably stored near each animal. Drinking water could have been piped into troughs, just as the Chinese have done using bamboo pipes for thousands of years. Self-feeders would have provided the food for a week at a time. Animal enclosures could have been sloped, with self-cleaning floors that emptied into manure gutters or pits. Or the animals could

## INSPIRED BY BIBLICAL ACCURACY

June 30th

have trampled their manure into pits through slatted floors. Even if as much as twelve tons of manure was produced daily, there were ways to handle the waste. Studies show that eight people could have fed and watered 16,000 creatures with no problem. While the year long stay on the Ark may not have been comfortable or easy, it was certainly do-able.

**A window shalt thou make to the ark, and in a cubit shalt thou finish it above; and the door of the ark shalt thou set in the side thereof; with lower, second, and third stories shalt thou make it.**

**- Genesis 6:16**

*Democracy never lasts long. It soon wastes, exhausts, and murders itself. There was never a democracy that did not commit suicide.”*

**- John Adams** (1735-1826)
Second President of the United States

*And GOD saw that the wickedness of man was great in the earth, and that every imagination of the thoughts of his heart was only evil continually.*

**- Genesis 6:5**

What causes an ice age? An ice age results when the winter's snows do not melt each year but are continually added to. It takes very special conditions to make an ice age. It may sound strange, but an ice age can only happen if the oceans heat up so that there is an increase in water evaporation and cooler continents that keep the snow from melting. What event in history would have lots and lots of warm water and continents that were cooler? The Flood of Noah's time. During the Flood the fountains of the deep burst open.

Widespread underwater volcanic eruptions spewed out great amounts of lava – heating ocean waters. These volcanoes also created unheard of quantities of volcanic dust; these particles reflected the sunlight back into space, making the continents colder. Rapid continental movement also would have added enormous amounts of heat to the oceans. These warmer oceans resulted in



increased evaporation and winds carried the moisture onto the cold continents. The cold continents caused the moisture in the air to condense and fall as snow. Snow on the ground did not melt during the summer because the volcanic dust blocked the sunlight. Each year the snow piled up deeper and deeper. Immediately after the Flood, ice sheets formed around the world in the higher latitudes such as Greenland and North America. As the earth settled down and, the volcanoes stopped erupting, the volcanic dust dissipated. Eventually the oceans cooled and the sun started to melt ice sheets. Creation scientists calculate the Ice Age to have lasted about 700 years – 500 years to build up and 200 years to melt down.

What causes an ice age? Very special conditions. An ice age needs lots and lots of warm water and cooler continents. What event in history would have created these special conditions? The Flood of Noah's time.

**Out of whose womb came the ice? and the hoary frost of heaven, who hath gendered it? - Job 38:29**

## Inspired by PALEONTOLOGY

July 2nd

Organic molecules rapidly degrade when exposed to oxygen or weathering forces. Even when deeply buried, organic materials cannot survive for tens of thousands of years. The fragile chemicals that make up our bodies simply "unzip" without the mechanisms of a living cell to protect them. Both biologists and geologists know that undecayed tissue could never survive for millions of years while buried in the ground. For this reason, biologists had never even thought of looking for undecayed organic tissue in fossilized bones. Evolution requires the assumption that the fossil record was not created by a recent flood but is hundreds of millions of years old. Thus, no-one looked for undecayed tissue within fossils. But this all changed in 2005 when undecayed blood vessels and ligament tissues were found in a Tyrannosaurus Rex bone.[1]

Since this discovery, undecayed tissue has been discovered throughout the fossil record.[2] A recent paper even demonstrated that remnants of organic tissue could be identified in skin impressions of a fossil which were supposed to have been buried 50 million years ago! It is becoming absolutely apparent these millions of years are simply a fantasy needed to explain life without God. The reality that even undecayed skin tissue can be found within fossils essentially proves that the fossils were buried quite recently.

**Ye blind guides, which strain at a gnat, and swallow a camel.**

**~ Matthew 23:24**

How much time do you spend thinking about bacteria, fungi, and viruses (collectively known as microbes)? If you are like most people – not much! Yet, where did these disease-causing microbes come from if God declared His creation "very good"?

When creation biologists view microbes from the creation perspective, they start with the biblical assumption that the original microbes were beneficial. Let's look at one microbe which is feared for the death it causes: cholera. Cholera causes severe intestinal illnesses in humans. It commonly comes from contaminated food or water. Cholera is caused by the bacterium Vibrio cholera, which makes a variety of poisons. However, most strains of Vibrio cholera grow harmlessly on almost all ocean shells and some fish. The purpose: to break down chitin (the stuff sea shells are made of) or the exoskeletons of shrimp, crabs and other sea creatures. If this microbe did not break down chitin, our oceans and beaches would be littered with billions of shells. With the chitin broken down, nutrients are returned back to the oceans. Cholera has a beneficial purpose. It was created "very good".

Just because something in nature causes harm, there is no reason to conclude that it was originally created to have this purpose. Death, disease, bloodshed, and evil exist because of the downward deterioration of creation once sin entered the world, not because everything was originally created this way. These



things are mankind's fault – not God's! The useful purposes of even deadly bacteria, such as cholera, clearly demonstrate this Biblical principle.

**The earth, O LORD, is full of thy mercy: teach me thy statutes.**

**~ Psalm 119:64**

According to NASA, "Hurricanes are the most powerful accumulations of energy on Earth". They are violent ocean-churning storms. Of what benefit are they?

- Hurricanes release a tremendous amount of rainfall. In 2003, Hurricane Isabella dropped 400 trillion tons of fresh water before blowing out.
- Hurricanes act as huge desalination factories bringing much needed fresh water to hot, drought-affected countries.
- Many regions of the world owe their harvest success to hurricanes. Hurricanes water the crops and replenish ground water.
- Hurricanes cool the ocean. The hot surface waters are replaced by deeper, cooler water moving upward after the removal of hot ocean water via evaporation. By using satellites to track the path of a hurricane, we can "see" the heat content of the ocean's surface water. The trailing pathway of a hurricane is considerably cooler after the storm passes. After Hurricane Fabian passed, the ocean waters were 10 degrees cooler.
- Hurricanes act as a safety valve and spread this extra heat out away from the tropical zones toward the poles. When the ocean waters evaporate, they remove the heat and release it into the atmosphere, often far from where the storm first picked it up.

Yes, hurricanes are powerful storms, but they do much more than harmful destruction.



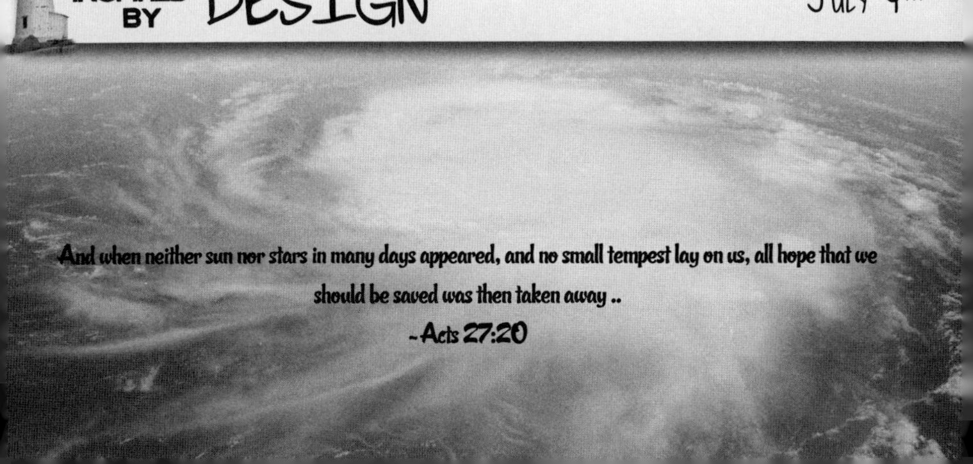

*And when neither sun nor stars in many days appeared, and no small tempest lay on us, all hope that we should be saved was then taken away ..*

*-Acts 27:20*

Truth, grounded in the word of God, is a powerful tool in the Christian arsenal. Truth forces those lost in the shifting sand of relativism to face reality. A person armed with truth can have a tremendous effect far beyond his individual influence.

July 5th

## CHRISTIAN TRUTH INSPIRED BY

Let's suppose all of the merchants within a town decided to work together to fix prices at an arbitrarily high level. In essence, they would be lying about the true value of the goods while justifying their actions. It would only take one honest merchant (armed with the truth and a biblical desire to treat others as himself) to end the price fixing. If a single merchant sold the same items at a lower price, the word would spread, and soon everyone would buy from him – forcing all of the dishonest merchants to bring their prices in line. The downside is that this honest merchant would find himself vilified by the other merchants.

In a similar way, those who promote the Bible's straightforward truth on both moral and scientific issues are vilified by those who wish to live a lie. Whether the topic is absolute honesty, sexual morality, the sanctity of life, or exposing the problems with evolution…those who dare to circle themselves with the belt of truth and speak the truth in love (using the word of God as their source of truth) are painted as old-fashioned, bigoted, outdated, judgmental, and ignorant. Yet just as with the honest merchant, even a single voice proclaiming the truth is like a light in a dark room. It can be seen by those searching for enlightenment and keeps the darkness at bay.

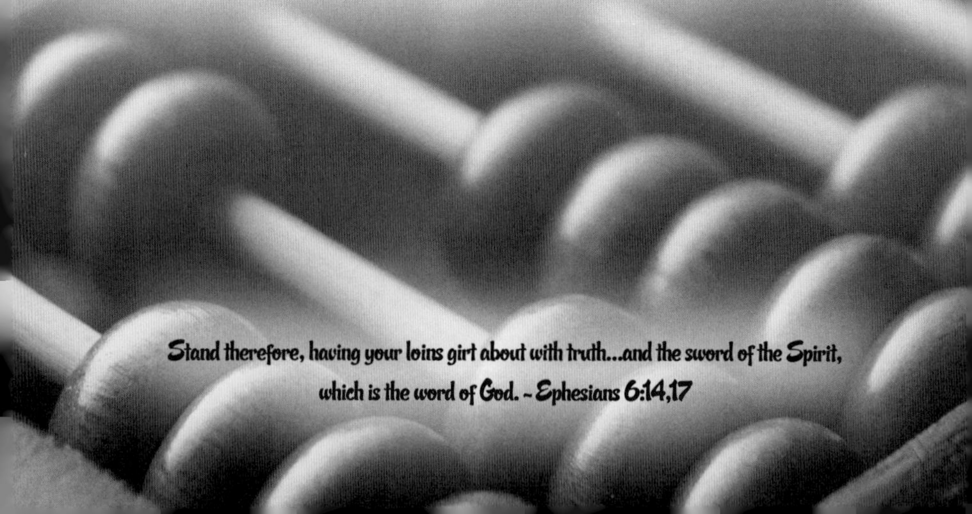

**Stand therefore, having your loins girt about with truth...and the sword of the Spirit, which is the word of God. - Ephesians 6:14,17**

Quick, go find an ant. Put the ant in a glass of water; be sure to keep pushing it down as it tries to climb out of the glass. Soon the ant will look like it has drowned; it will be lying at the bottom of the glass of water. Leave the ant there for a few minutes. Now remove the ant and pour some salt on it (the salt helps the ant dry out more quickly). Shortly, the ant will begin to move about as if nothing happened.

How does the ant do this, and why does it not drown? Ants and other insects do not breathe like we do; they breathe through tubes in their bodies called spiracles. All over their abdomens are little holes that let air into the bodies. When an insect is faced with heavy rains or floods, they just close off these spiracles. They are able to survive for weeks by closing the spiracles.

Now consider this question: Did Noah take insects on board the ark? I suspect not. He did not have to because they would have survived the flood by closing their spiracles. Genesis 7:22 states, "Everything on dry land that had the breath of life in its nostrils died." Insects do not have nostrils, and they could have survived quite well outside of the Ark. All other types of land animals, those that breathed through their nostrils, died.

Even if Noah had collected and put representative insects into cages, one million different kinds would only fill 12 railroad cars. This is miniscule compared to the size of the entire Ark (which had a size equivalent to 522 railroad cars). Insects could and did survive Noah's flood.

**All in whose nostrils was the breath of life, of all that was in the dry land, died.**

**- Genesis 7:22**

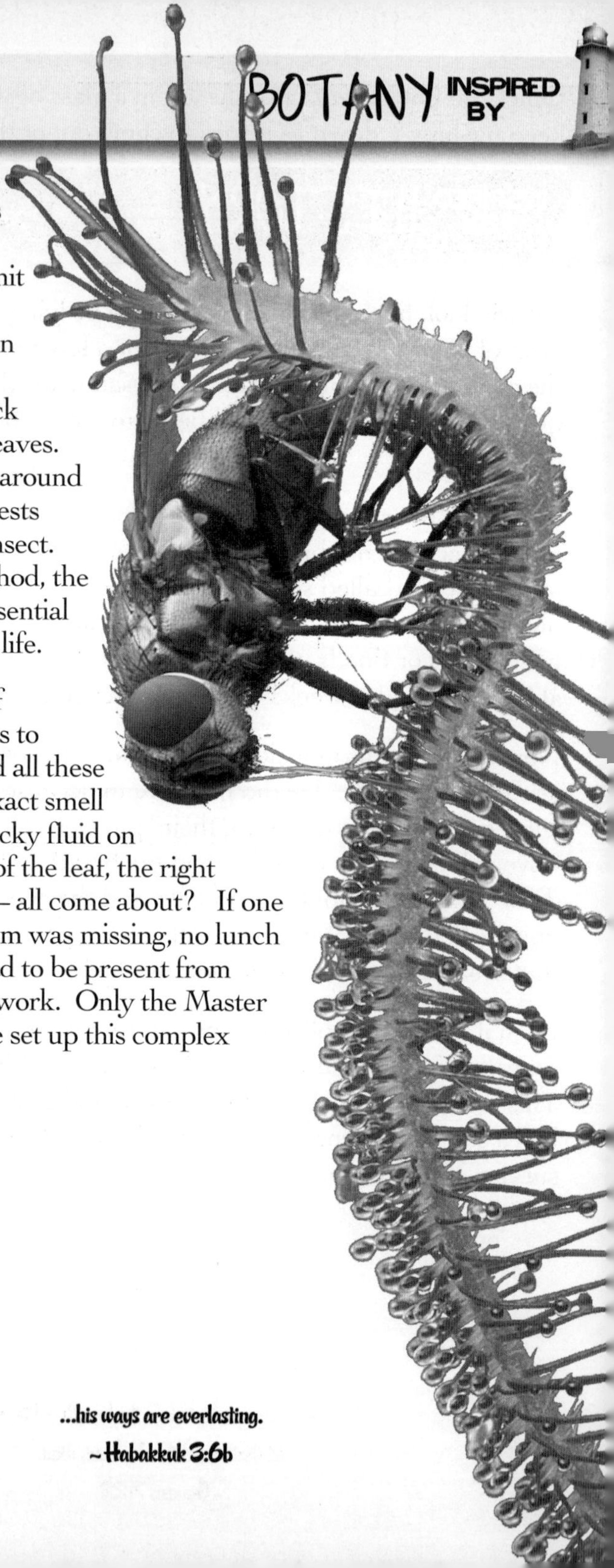

The sundew plant traps insects with its long red hairs. The tips of these hairs emit a smell that entices insects to visit. Upon an insect's arrival, it gets stuck to the thick fluid on one of the leaves. Then the leaf closes around the insect, and it digests nutrients from the insect. By this hunting method, the sundew gains the essential proteins it needs for life.

What an odd way of hunting; the food has to come to it! How did all these mechanisms – the exact smell for attraction, the sticky fluid on the leaf, the closing of the leaf, the right digestion chemicals – all come about? If one part of the mechanism was missing, no lunch for the plant! All had to be present from the start in order to work. Only the Master Designer could have set up this complex method of hunting.

***...his ways are everlasting.***

**~Habakkuk 3:6b**

Our universe is truly a spectacular place. The heavens are enormous because God made them to reflect his character: beauty, power, wisdom, and creativity. Psalm 19:1 tells us that "The heavens declare the glory of God; the skies proclaim the work of his hands."

Consider what the heavens show us:

- **Nebulas:** The Hourglass and Rosette are breathtaking clouds of hot gas and dust producing spectacular beauty. This beauty proclaims God's creativity.
- **Globular Star Clusters**: These are tightly knit groups of hundreds of thousands to millions of stars. They appear as



fuzzy spots of light through a beginner's telescope. The number of stars viewed in one spot is staggering. This shows God's power.
- **Elliptical and Spiral Galaxies**: These are but two shapes of the hundreds of billions of galaxies in the universe. Galaxies can measure from a few thousand light-years across to more than half a million light-years. Light years are a form of distance measurement, not a measure of time. This display of such great expanses declares God's might and glory.
- **Saturn:** The sixth planet in our solar system is recognized by its stunning rings of ice and dust. Saturn, with its rings, displays God's creativity and beauty.

The universe is a canvas upon which God declares His glory.

**The Lord by wisdom hath founded the earth;**
**by understanding hath he established the heavens.**
**~ Proverbs 3:19**

Bird beaks are made of hard keratin, the same material as our fingernails. These beaks grow continually throughout the life of a bird. England's Great Tits can change their beak shape effortlessly twice a year (thinner for summer insects and thicker for winter nuts). Oystercatchers were observed to change the shape of their beak – sharp or blunt – according to their food preference. It has been found that Hawaiian Honey Creepers will rapidly change their bill shape to accommodate differing flowers. This brings us to Darwin's Galapagos finches. Darwin used finch beaks to "prove" evolution. He saw minor changes within the beaks and then extrapolated. Darwin thought that given enough



time, anything could turn into something else. Darwin did not know about the incredible complexity of the cell or genetics; to his generation, the cell was just a blob of "protoplasm." Since then, scientists have been studying these Galapagos finches and have found that their bills change shape within the egg while the beak is forming as the food source varies. The bird's bill shape is now thought to be governed by a protein called BMP4. BMP4 signals the molecule that controls the formation of bone. The amount of protein that is present causes the finch embryo to produce either a thick stout beak or a narrow pointed beak. These changes are not the result of mutations but design within the organism. The activation of the protein BMP4 allows the finch to adapt to the changing environment – depending on whether the environment is drier or wetter. This is God's "programmed filling" in action. It allows the finch to fit the changing environment.

But the finch is still a finch, it simply has variations in its beak size. God created the bird kind with variation within each kind. With Darwin's Galapagos finches, we see just that variety within the finch kind.

***...and every winged fowl after his kind.***

***~ Genesis 1:21***

A polystrate fossil cuts across many geological layers. Poly means "many" and stratum means "layer", hence "many layers". The eroding cliffs of Joggins, Nova Scotia are famous for their abundance of polystrate or fossilized trees that cut through many layers. The evolutionists' explanation for a tree to be polystrate is that a tree was slowly covered with sediment over millions of years. But does this explanation really make sense? Wouldn't the tree rot away during all that time?

Polystrate fossils are found on a worldwide scale. Specimen Ridge in Yellowstone has numerous fossilized trees cutting through many rock layers. Ginkgo Petrified Forest State Park in Washington has multiple examples of trees that are also polystrate. In Lompoc, California, an upright whale fossil cuts up through many layers. Wouldn't the whale have rotted if it had taken millions of years to be covered? In Peru, 346 well preserved whales and other animals have been uncovered. These whales were also polystrate. Coal mines frequently contain polystrate trees.

The one-year Genesis Flood perfectly explains such geological observations. Fossilized trees and whales are found penetrating through multiple layers of sediment because these layers were formed rapidly, not slowly over millions of years. Polystrate fossils point to rapid burial and the reality of Noah's flood. The science of paleontology is filled with dead creatures whose orientations in the rock layers testify to this flood event. God has left us a world filled with polystrate fossils as evidence for a worldwide Flood.

**Thou didst cleave the fountain and the flood...**
**~Psalm 74:15a**

It was World War II, and the British Royal Navy ships were experiencing unexplained damage to their propellers. Physicists worked out the problem: it was cavitation bubbles. Cavitation bubbles occur when turbulent waters cause tiny bubbles to grow and then collapse. These tiny bubbles can rise to temperatures of 27,000 degrees F. (as hot as a star's surface). The result is great damage where the bubbles burst.

Little wonder that this same cavitation mechanism cut through solid concrete dam tunnels at Glen Canyon Dam just north of the Grand Canyon in 1983. Unexpected rains put pressure on the dam's spillway causing slight rumblings and vibrations. One of the spillway



tunnel's portals erupted with jets of water containing debris of concrete, rebar and rock (one boulder measured 10 feet by 15 feet). Upon inspection, the tunnel had a new hole, 50 feet by 135 feet. The high speed water had cut through the reinforced concrete and sandstone. Cavitation had done its work.

Now imagine the destructive power of rushing waters as they poured off the continents at the end of the Global Flood. We can see the leftover signs of fast flowing water scouring the land – cavitation in the steep-sided canyon, gorges and ravines of the world.

**Which removeth the mountains, and they know not: which overturneth them in his anger. ~ Job 9:5**

# INSPIRED BY MICROBIOLOGY

July 12th

Umbrellas are not just for rain; they also provide shade from the sun. The sun gives off deadly ultraviolet rays which would kill every one of us if we did not have protection. Sunscreen helps, but God provided us with microscopic umbrellas, in fact, millions of them. These umbrellas are melanin. Melanin is what gives us our skin color. Melanin is produced in melanocytes. Once melanin is made, it is packaged into tiny granules which are shipped to the tips of the melanocytes. The skin, or epidermal cells, then "bite off" the tips of these projections.

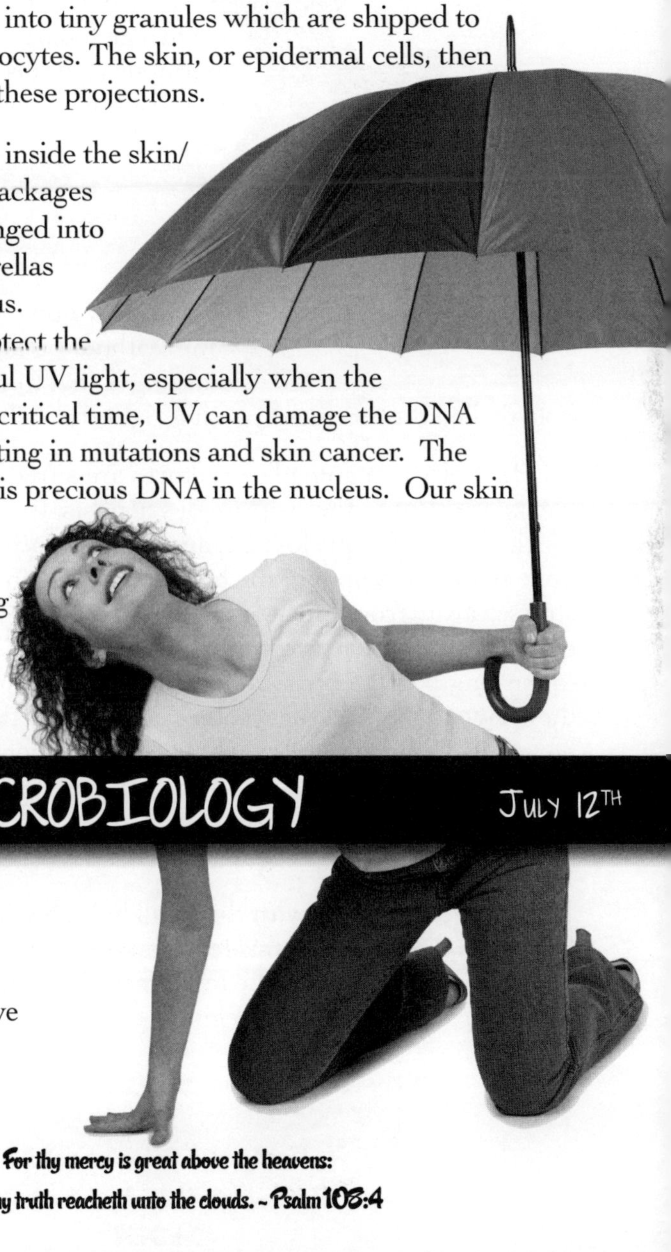

Once the package is inside the skin/epidermal cell, the packages are moved and arranged into dark "caps" or umbrellas over the cell's nucleus. These umbrellas protect the nucleus from harmful UV light, especially when the cell divides. At this critical time, UV can damage the DNA in the nucleus, resulting in mutations and skin cancer. The umbrellas protect this precious DNA in the nucleus. Our skin contains hundreds of millions of tiny umbrellas containing melanin to protect us from overdoses of the sun's deadly radiation. God's protection is more than skin deep. What a great God we have!

**For thy mercy is great above the heavens: and thy truth reacheth unto the clouds. ~ Psalm 108:4**

Did you notice the winning horse's shin bone (or metacarpal) the last time you watched a horse race? This tiny shin bone is located just below his front knee, supporting the horse's whole weight, even when galloping. Think about the enormous load this bone (the size of our wrist bone) experiences as a horse



thunders along. How does it do this without breaking? Even more amazing is that this tiny bone has a hole (called the foramen) passing right through it.

Similar holes in manmade structures are a frequent source of weakness. But to the surprise of scientists, this is not a problem for the horse. Upon close examination, scientists have found that bone tissue surrounding the hole is arranged in a way that directs stresses away from the hole toward stronger regions of the leg bone. Also, the hole is elliptical with the ellipse's long axis parallel to the length of the bone. If a race horse fractures his metacarpal, they have found that the fracture is seldom at the opening. The horse's leg bone design is such that the foramen is actually a source of strength. This discovery may solve some age-old engineering problems, such as how best to put holes in material such as portholes in ships.

The horse's shin bone with the small hole shows exquisite design. Evolution did not make this foramen, for when we look at the wide variety of horses in the fossil record, the foramen opening is always present. From the beginning, God designed the horse with a foramen.

**Hast thou given the horse strength?**
**Hast thou clothed his neck with thunder?**
**~ Job 39:19**

A strong evidence for the rapid formation of the geology of our planet is the widespread existence of folded sediment throughout the mountain chains of the world. Three-fourths of the surface of the Earth is made up of sedimentary rocks. Sedimentary rocks are made up of individual particles or sediments compressed and hardened to form rocks such as shale, limestone, slate, mudstone, coal, and sandstone. Sedimentary rocks are generally formed underwater. Surprisingly, even the highest mountains on Earth, jutting an incredible 29,000 feet above sea level, are covered with sedimentary rocks.

Geologists find these sedimentary layers throughout the mountain ranges of the world. Surprisingly, many of the mountain's sedimentary layers show signs of being bent, not fractured or broken. How can rocks be bent and not broken? If mountain ranges were pushed up slowly over millions of years after sediment had been laid down,



the rocks would have broken like a stack of China plates being pushed up in different places. Unbroken, bent sediment layers could only happen if the mountains were pushed up rapidly while the sediment layers were still soft and pliable. Heated rock can be bent, but the structure of the rock changes and these bent sediments show **no** sign of this kind of structural change.

The Bible explains the evidence best. The worldwide flood left thousands of feet of unconsolidated and re-arranged sediment across the surface of the planet. During the Flood, there would have been rapid continental movements causing the mountains to be uplifted. The recently laid down flood sediments would have been soft and pliable when the uplift took place. This would have resulted in the folded and bent sedimentary rocks within the mountain chains. Thus finding folded and bent sedimentary rock layers is not only to be expected but predicted by this Biblical model because the rock layers could have been soft and pliable when the uplift took place.

**They go up by the mountains; they go down by the valleys unto the place which thou hast founded for them. ~ Psalm 104:8**

Have you considered the heart of a baby in the womb? How does a baby breathe when emmersed in a sack of water? Why doesn't blood circulate through a baby's lungs while the baby is inside of the mother? The blood does not need to circulate to the lungs because the baby receives its oxygen from the mom's placenta via the umbilical cord. Yet the heart is a vital organ that must function from the beginning. So what changes in the baby's heart at birth?

## JULY 15TH — ANATOMY INSPIRED BY

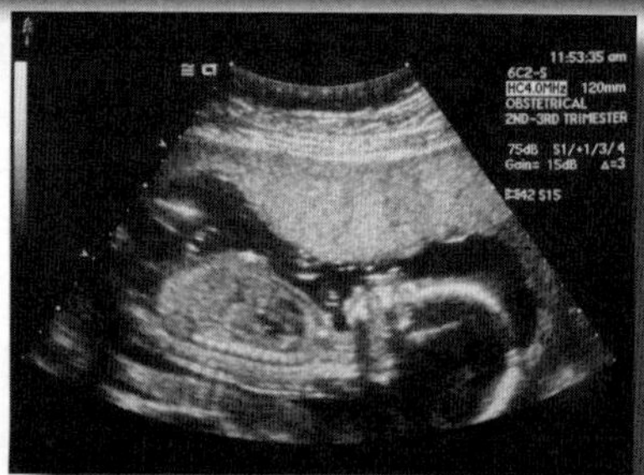


The unborn baby's lungs are collapsed; therefore, the right side of the heart does not need to send blood to the lungs. There are two short-cuts in the unborn baby's heart that enable the blood to bypass the lungs. One is called the "foramen ovale" and the other is "ductus arterioseus". The foramen ovale is a hole between the right and left side of the heart. Blood flows from the right side directly through the wall into the left side. An unborn baby has a hole in its heart. At birth, as the baby makes its first cry, oxygen sweeps in, the lungs are inflated and the flap slams shut, closing the hole. Blood now starts circulating into the lungs to get oxygen.

The other lung bypass is the ductus arterioseus. This blood vessel routes the blood directly into the aorta. When the baby is born, the body makes chemicals that cause this blood vessel to instantly close. The blood then begins moving through the lungs to absorb oxygen.

How do evolutionists explain this complex and perfectly designed set of engineering marvels? What if no flap developed? Holes in the heart are not a good thing. What if the extra bypass blood vessel did not close down? Instant death! We can breathe a sigh of relief that God designed this marvelous heart to function wonderfully before birth and after birth—without a moment's delay.

**For thou hast possessed my reins: thou hast covered me in my mother's womb.**
**14I will praise thee; for I am fearfully and wonderfully made: marvellous are thy works; and that my soul knoweth right well. ~ Psalm 139:13,14**

Even though God is primarily concerned about your being in heaven with Him, He also cares about your health while you live out your life on Earth. Even the details of handling human waste are spelled out within Scripture – for our own good. The Israelites spent 40 years wandering in the wilderness, and it is estimated that there were several million individuals involved. Imagine a city of two million people without any sanitation or sewage handling systems – it would be a disaster. So God made sure the people knew how to handle this problem.

God commanded His people to have a place outside of camp where they could relieve themselves. They were to carry a shovel, so they could dig a hole and cover up their waste. By keeping the waste outside of the camp, it reduced the possibility of getting intestinal diseases such as cholera, amebic dysentery, and E. coli.

## INSPIRED BY BIBLICAL ACCURACY

JULY 16TH

This was written some 3,500 years ago, yet up until World War I, more soldiers died from diseases caused by not isolating human waste than from war itself. If only they had read their Bible and applied it as literal truth. God cares not just about our salvation but about our earthly life too.

***You shall also have a place outside the camp and go out there, and you shall have a spade among your tools, and it shall be when you sit down outside, you shall dig with it and shall turn to cover up your excrement.***

***~ Deuteronomy 23:12-13***

The ferocious piranha is a carnivore, which hunts in schools of 30 or more, preying on animals such as baby birds that fall out of their nests and into the river. They have teeth so sharp that the Indians of South America use them as razors. When God created the world it was perfect; there was no death. It was not until Adam sinned that death entered the world. So what did these piranhas eat before the Fall?

Scientists have now found evidence that the ancestors of the piranha once ate plants. The pacu fish are related to the piranha;



recent DNA analysis shows no significant genetic difference between the two species. In fact, the pacu fish are often misidentified with the piranha. Yet pacu fish are mainly vegetarian; they eat aquatic plants and fruit that have fallen in the water. In the same way, piranhas will, on occasion, eat plant matter, even though they are primarily carnivores. So why did some of the piranha/pacu fish become carnivores and others vegetarians?

After man sinned, mutations became a reality. These mutations may have led some of the original vegetarian kind to start scavenging dead animals and fish, which they found floating in the water. The pacu and the piranha are a variation of a kind (just like we have lots of varieties/breeds of dogs, but they are still dogs). So what did all of the ancestors of the piranhas eat in the beginning? Plants!

**And to every beast of the earth, and to every fowl of the air, and to every thing that creepeth upon the earth, wherein there is life, I have given every green herb for meat: and it was so. – Genesis 1:30**

## INSPIRED BY HISTORY — July 18th

During one of the manned missions to the moon in the 1970's, there was an explosion on the Apollo 13 space capsule. After the explosion, Apollo 13 radioed home one of the great understatements of history, "Houston, we have a problem." Over the next few days, NASA and the astronauts solved nearly every problem thrown at them. In addition, President Nixon publicly asked the entire nation to pray for their safe return. But one problem could not be solved; the lunar module was getting cold, almost freezing (38°F). The astronauts were alone, freezing, without warm clothes, or any source of heat, and at least three days travel from Earth. How did the astronauts survive?

What NASA couldn't solve, God had already provided for. God knows we will meet with unexpected dangers and has designed our bodies with backup systems for such emergencies. In the case of extreme cold, the brain's hypothalamus is ready. The first defense is to generate heat, so the astronauts started shivering. The second defense is to conserve body heat. The blood vessels just below the skin's surface begin to constrict, keeping the blood deeper and warmer as it circulates. As the cold increases, our heart rate and digestion slow (especially useful when all there was to eat were blocks of frozen food!) Next, the brain triggers the blood to concentrate around the heart and brain, protecting these vital organs. Fingers, toes and other extremities are left to the cold.

As the astronauts' bodies became colder, their nervous systems slowed and with this, clear thinking became more difficult. The astronauts struggled to understand and recall what Mission Control had told them. By the time they entered Earth's atmosphere, their brains were in survival mode. Because their capsule was on "autopilot", their Command Module splashed safely into the ocean.

How did the astronauts survive? Our brains are wired for the extremes; God has equipped our bodies with sophisticated contingency plans to help us survive in life-threatening situations.

**God is our refuge and strength, a very present help in trouble.**

**- Psalm 46:1**

In addition to the upper and lower eyelids, birds have a third eyelid. This transparent eyelid (nictitating membrane) sits in the corner of the eye and sweeps backward. The purpose of this membrane is to clean and moisten the eye while still allowing the bird to see. Imagine a bird without this "windshield" flying on a dusty, windy day. Closed eyelids would cause momentary loss of vision – this would not be good when plummeting earthward for a rodent or darting in and out around branches. Imagine falcons diving at great speeds and not having been design with this third eyelid; their eyes would immediately dry out. This third eyelid allows protection



and vision at the same time. As for ducks, they just put on their "goggles" in order to have clear, undistorted vision while swimming underwater. We have a Master Designer who has concerns even for a bird's eyes.

**I know all the fowls of the mountains: and the wild beasts of the field are mine.**

**~ Psalm 50:11**

How can we know if information is accurate or biased? New discoveries, studies and information are available every day, but all too often they have been interpreted for us from an evolutionary viewpoint.

## INSPIRED BY CHRISTIAN TRUTH — JULY 20TH

With the availability of the Internet, it is now easier to find the original sources of research, archaeological discoveries, etc. We need to look for biases, test/verify the interpretations, and draw our own conclusions without the evolutionary spin. Sometimes the original premise and assumptions of a research project are based on evolutionary theory. Watch for these biases. Look at the data, and see what it actually shows. You will find that accurately interpreted data will ALWAYS be in agreement with the Bible. If you see something that seems contrary to the Bible, it has been shown over and over again that either there is a misinterpretation or some aspect has not yet been discovered. A recent example of this is our current understanding of genetic deterioration as explained by Dr. John C. Sanford.[1] His work shows that the information in the DNA of all creatures is being rapidly corrupted – essentially proving that no creature upon Earth could be millions of years old.

Numerous websites review recent scientific discoveries without an evolutionary bias. Links to many of these sites can be found at www.searchforthetruth.net.

**Beware lest any man spoil you through philosophy and vain deceit, after the tradition of men, after the rudiments of the world, and not after Christ.**

**~ Colossians 2:8**

Have you considered the clouds? Clouds both heat and cool the Earth. When clouds cover the sky, they help keep the Earth warm, like a blanket, by reducing the amount of surface heat radiating into space. There have been estimates that clouds warm the Earth by an average of 9°F. Clouds can also keep the planet cool by not allowing the sunlight through but reflecting the sunlight back into space. Without the sunlight getting through, it is estimated that surface temperatures are reduced by as much as 22°F on average across the planet.

Clouds act like a thermostat preventing runaway temperatures. In spite of dire warnings of global warming, what we actually observe is a balanced temperature upon our planet due to the action of clouds. As the Earth warms, it produces more clouds through evaporation that then prevent the sunlight from warming the Earth further. As the Earth cools, fewer clouds are produced, allowing more sunlight to reach the surface and heat the planet. Clouds both heat and cool the Earth!

God has built into creation many such complex processes – which we are only beginning to understand.

## July 21st



**Also can any understand the spreadings of the clouds, or the noise of his tabernacle? ~ Job 36:29**

It takes very special conditions to make sedimentary rocks such as limestone, sandstone, and shale. Sedimentary rock is made up of small bits of particles from precipitated lime, clay, silt, and sand. How do these loose sediments turn into a rock? Sedimentary rocks form the same way concrete hardens. After all, concrete is just artificial rock! Concrete manufacturers basically break down big rocks into powdered rock. To make concrete, water, sand, and pebbles are added to the powdered mix until just the right amount of each is mixed together. The two most common rock cements are calcium carbonate and silica. Calcium carbonate is the lime in limestone while silica rapidly absorbs water to form hard crystals.

Sedimentary rocks cover three-fourths of the Earth's surface. What event in history would have pulverized enough pre-existing rock and sediment to cover much of the planet's surface with a



concrete-like mixture? The Flood of Noah's day. As we read in Genesis 7:11-12, "All the fountains of the great deep broke up and the windows of heaven were opened. And the rain was upon the Earth forty days and forty nights..." The entire Earth became covered with water, "And the waters prevailed exceedingly upon the Earth; and all the high hills, that were under the whole heaven, were covered....and the mountains were covered." – Genesis 7:19-20. This was no small local event!

Now imagine this year-long flood with tsunamis and flood waters breaking up and eroding the Earth's rocks. Mix it all up, and what you have is an Earth-sized cement mixer. At the end of the Flood, the mountains rose up, and the flood waters rushed off the continents. Now the land had time to "cure", its cement became hardened into rock layers, and 'voila' – three-fourths of the Earth was covered with sedimentary rocks.

**Then the channels of waters were seen, and the foundations of the world were discovered at thy rebuke, O LORD, at the blast of the breath of thy nostrils. ~ Psalm 18:15**

The primary problem with Darwin's theory of evolution is not *survival* of the fittest, but *arrival* of the fittest. Darwin was never able to explain the origin of information needed to create new forms of life. Nor has anyone since the time of Darwin. Darwin's natural selection can create nothing. Living organisms have the pre-programmed traits to do better or worse within various environments. Those with the traits to do better reproduce in larger numbers and eventually become the main population for that environment, while those who do worse end up dying off. Nothing "new" is created.



It is true that Darwin presented voluminous support for natural selection in *Origin of Species*, but "ironically never explains where any new species came from".[1] The best Darwin was able to do was refer to a vague force he called pangenesis – an idea indistinguishable to Lamarckism (the disproven belief that learned characteristics are passed on to offspring), e.g. if I developed a muscular right tennis arm, my children would be born with this trait.

The best that modern evolution believers have come up with for explaining "the arrival of the fittest" is the belief that random mistakes to the DNA code (mutations) drive the upward advancement of life. Yet, no evolutionist hopes that their baby is born with a new form of birth defect. That is exactly what a mutation is – a birth defect. If mutations drive evolution forward, why not hope for birth defects in your children? Whenever evolutionary beliefs run into the wall of reality, evolution loses; the inconsistencies of even the most ardent evolution believers become readily apparent.

**The tongue of the wise useth knowledge aright: but the mouth of fools poureth out foolishness. ~ Proverbs 15:2**

There is a unity of teaching in the Bible that is startling in light of being written:

- By 40 different people: soldiers, kings, shepherds, statesmen and common men.
- By men living in different places: cities, deserts, palaces, and tiny towns.
- At different times: periods of war and peace, times of prosperity and great need.
- In different moods: feelings of great joy and great sorrow, moments of fear and times of security.
- From three different continents: in Egypt (Africa), in Italy (Europe), in Judea (Asia)
- In three different languages: Hebrew, Greek and Aramaic
- About many controversial topics: history, science, philosophy, geology, cosmology, the nature of God, the role of government, the nature of mankind, moral issues, biology, political alliances, religious ceremonies.

## INSPIRED BY BIBLICAL ACCURACY

July 24th

And yet the Bible is not a mix of contradictory ideas. It reads like one story with harmony and continuity from Genesis to Revelation. It truly is the inspired word of a Holy God.

**All scripture is given by inspiration of God, and is profitable for doctrine, for reproof, for correction, for instruction in righteousness: ~ 2 Timothy 3:16**

Each fall, more than 100 million monarchs migrate from all over North America to 12 small groves of firs in a 60 by 70 mile area on a mountain range west of Mexico City. And they do this without ever having made the trip before!

Let's follow their life cycle process through one calendar year:



- The monarchs leave Mexico in March – arriving in North America to mate and lay eggs. These eggs hatch, grow into butterflies, mate, lay eggs and die – all within about a month.
- Subsequent generations keep moving north, mating, laying eggs and dying.
- In August/September, the great-great grandchildren of the original butterflies that left Mexico are born. This generation must make the migration back to Mexico – a journey of up to 3,000 miles.
- The monarchs that spend the winter in Mexico live for about 8 months while the others live only 1-2 months. The entire round-trip migration of monarchs takes place over 3-5 generations.

How do they know where to go? Evolution says this all happened by chance; how could it have? There are too many steps. God set this up from the beginning to show his great power. God has bound the information into the genetic code of each butterfly so that each "knows" what stage of the migration cycle it has to fulfill. This information is passed on to each generation. This shouts design.

**Be thou exalted, O God, above the heavens;**
**let thy glory be above all the earth.**
**- Psalm 57:5**

## INSPIRED BY BIOLOGY

JULY 26TH

Far more incredible than its migration, is the monarch's metamorphosis. Everything about this creature is unexplained by evolution.

- How does a caterpillar make a chrysalis and dissolve itself into jelly…and come out a butterfly?
- How could a creature with 6 simple eyes, which see in black and white…create two new eyes, each made up of 6,000 individual lenses with the ability to see in color?
- How could a creature with a chewing mouth that eats only leaves…develop a long tongue that sucks only nectar?
- How could a creature with 16 legs become jelly…and produce completely differently shaped legs (6) and wings (4)?
- How could this have happened by chance…when any flaw along the way would have produced death?

God, the master engineer, designed this incredible system; whereby, the caterpillar seals itself into a chrysalis, dissolves itself into jelly, and emerges as a butterfly. As you see the monarch's migration, be amazed by its 3,000 mile flight…but be equally amazed at how it transformed itself into a butterfly!

**Therefore if any man be in Christ, he is a new creature: old things are passed away; behold, all things are become new. – 2 Corinthians 5:17**

Jupiter is a huge gas planet, more than 300 times the mass of Earth. This "king planet" acts like a giant vacuum cleaner in space, protecting Earth from comet and solar system debris impacts. This was illustrated in July of 1994 when the Comet Shoemaker-Levy 9 collided into Jupiter. As Jupiter's



tremendous gravitational pull drew in this comet toward its surface, the comet broke into fragments. These fragments then collided with Jupiter in a spectacular collision, which was captured on film.

Uranus and Saturn also provide a gravitational pull that "vacuums up" debris. In addition, planets within the inner solar system protect us from being bombarded by asteroids from the asteroid belt. Mars is our first line of defense, being at the edge of the asteroid belt. It takes a lot of hits for us, as does Venus. One look at the moon's surface, full of impact craters, shows what a rough environment outer space really is – many of these asteroids would have impacted on Earth had the moon not intercepted them for us.

God has provided just the right amount of protection for the Earth by placing the planets in the right position with the right amount of gravitational pull. A coincidence you say? Hardly. The more we study the heavens, the more we see the hand of protection from our all-powerful God.

**Thy mercy, O Lord, is in the heavens... ~ Psalm 36:5a**

Each human is absolutely unique and known in an intimate way by his Maker. Not only does God know the number of hairs on our head, but He knows everything else about us. There is nothing He has designed and nothing we are experiencing which is not apparent to our Maker.

One example of our uniqueness is our iris. Every person's iris is a unique combination of blue-green-brown color pigments

## INSPIRED BY ANATOMY

JULY 28TH

arranged in a specific pattern. Look closely at your iris, and you will see a number of star-like patterns. Each person's iris actually has at least 266 identifiable measurement points unique to any given individual. By comparison, our fingerprints have only 35 measureable points. The odds of one person having exactly the same iris as any other person is 1 out of $10^{78}$. This probability is FAR less likely than randomly selecting a single special electron from a random mixture of every electron in the entire universe, i.e. an impossibility.

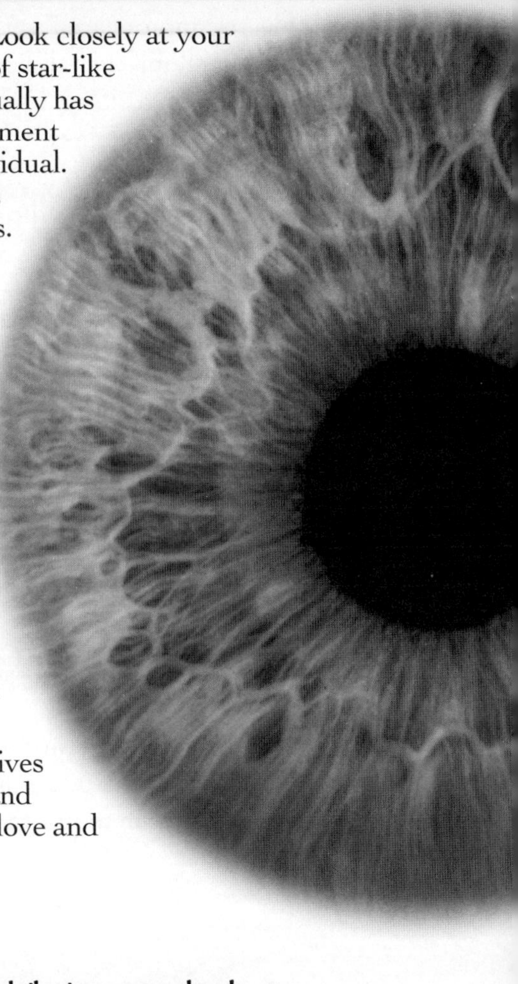

Every person that God created is totally, uniquely, and individually special. God knows everything about you and loves you more than you can possibly imagine. Yet, without freedom, there can be no love, so Jesus gives you the freedom to love, obey, and worship Him…or to reject His love and live for yourself.

**But as it is written, Eye hath not seen, nor ear heard, neither have entered into the heart of man, the things which God hath prepared for them that love him. ~1 Corinthians 2:9**

Evinrude, the name is proudly printed on the outside of an outboard motor. We know that it took teams of engineers to design and make this motor – all so that we can move through the water in our boats. We can see the same marvelous motor engineering in miniature, in the spinning flagellum (tiny hair-like structure) of bacteria such as E. coli. How does a bacteria move through its watery environment? It uses a motor-driven propeller!

This flagellum can stop, start, change directions, and go in reverse while spinning its propeller at up to 100,000 rpm (revolutions per minute). Bacterial motors are almost 100% efficient while man-made electric motors are 75-95% efficient. When scientists examined this flagellum, they discovered the same parts that we see in a motor: a rotor, a stator, O-rings, bushings and a drive shaft. Forty different proteins are used to assemble this little motor. These flagellum motors are so tiny that eight million of them would fit in the width of human hair - the ultimate in miniaturization!

Evolutionists say this all came about by accident over millions of years. However, if any part is missing, the flagellum motor could not work. If your outboard motor did not have a rotor, O-rings, bushing, etc., or if they were in the wrong place, the motor would not work. All the parts had to be there right from the beginning in order for the motor to perform. The same applies to the miniature motorized flagellum; all the parts had to be there from the beginning and in the right order.



So this summer, as you putter around the lake with your outboard motor, think of the bacteria puttering around in its watery environment with its own built-in motor that was designed by God.

**So when they had rowed about five and twenty or thirty furlongs, they see Jesus walking on the sea, and drawing nigh unto the ship: and they were afraid. ~ John 6:19**

The emperor penguins spend their entire lives in the frigid Antarctic. Each year the mother lays one egg on the feet of the father, and father covers the egg with a flap of skin. Now the mother leaves and goes fishing for two months. Father is literally left with egg on his feet! For two months, these dads are left to brave the frigid winter of icy temperatures, cruel winds, and blinding storms (at times dropping to -100°F). How do they stay warm? The emperors huddle together and take turns moving to the inside of the group. Once they have had a chance to warm up, they shuffle to the colder outside edge of the huddle.



During this time, each dad loses a lot of weight. You may have heard favorable things about the Atkins diet or the LA diet, but you would not want to try the penguin diet. He loses 2/3 of his body weight by not eating for two months! When the baby hatches, Mom has ten days to get home before Junior starves. How does she know when to show up? She arrives right on time every year, having waddled some 50 miles with a belly full of fish for the baby. Dad now leaves to go fishing.

Why do these crazy penguins choose the dead of winter to lay their eggs? Because when the baby is ready to start his swimming lessons in the spring, the weather has warmed, some of the ice has melted, and the distance to the shoreline is the shortest, bringing the open waters closer to the nesting sites. When the chicks are old enough to take to the seas and fish on their own, they have the shortest distance to travel.

God in His wisdom cares for all his creatures. He has put within the emperor penguins the perfect instincts needed for their survival. When we see instincts, we know there must be an instinct maker… and that is Jesus!

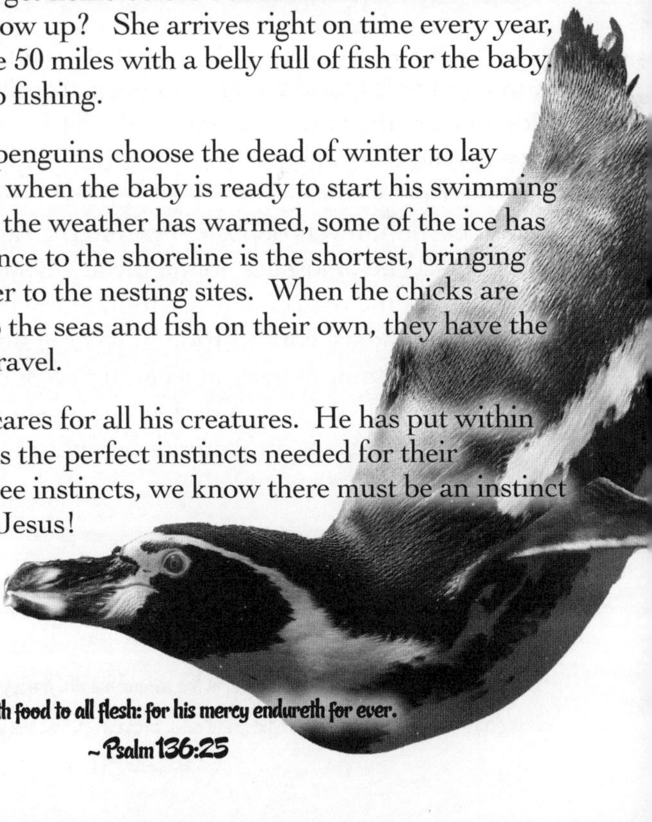

**Who giveth food to all flesh: for his mercy endureth for ever.**

**~Psalm 136:25**

Have you considered what killed the dinosaurs? Evolutionists will tell you that 65 million years ago the dinosaurs were killed off by a large asteroid or comet that hit our planet. With great authority they say that this impact left its evidence with the "K-T extinction event". The K stands for the Cretaceous layer of rocks and the T stand for the Tertiary rock layer. Dinosaur bones appear only on the Cretaceous side of the geological column. When scientists examine this K-T boundary, they discovered that the rocks in this boundary layer had an unexpectedly high concentration of iridium. They incorrectly assumed that the only possible source of this iridium was from outer space – hence the asteroid theory. Everyone agrees most dinosaurs were killed off by some event, but is an asteroid the only possibility?

It has been discovered that volcanic eruptions also produce iridium. The iridium in the K-T rock boundary layer did not have to come



from outer space. During the Flood of Noah, an unprecedented number of volcanoes would have been erupting, thus bringing up iridium from the bowels of the Earth. The watery flood would have spread the iridium across the globe. Apparently, at this particular point during the complex flood event, a high concentration of iridium was deposited within the rapidly accumulating sediments. The dinosaurs not on the ark would have died in this flood; hence, the reason they are found only in the cretaceous layer. The K-T boundary with its iridium layer is simply a confirmation of massive volcanic activity at a certain point during the worldwide flood. The evidence makes perfect sense when we start with the Bible and assume it is an accurate account of true history of our planet.

**In the six hundredth year of Noah's life, in the second month, the seventeenth day of the month, the same day were all the fountains of the great deep broken up, and the windows of heaven were opened.**
**~ Genesis 7:11**

# AUGUST

*"Give me five minutes with a person's checkbook, and I will tell you where their heart is."*

**- Billy Graham** (1918-present)
American evangelist, author, Christian leader

*For where your treasure is, there will your heart be also.*

**- Matthew 6:21**

Global warming, aka "climate change", is one of the big issues of the twenty-first century. In the 1970's, there were dire reports predicting a coming ice age. The 1980's saw common warnings of massive human starvation and war due to overpopulation. In the

August 1st

## CHRISTIAN TRUTH INSPIRED BY

1990's, it was the assurance that we would soon run out of energy. For those seeking to control people and maintain power, there is nothing like a good crisis to solidify funding!

There is much evidence that any recent warming trend is a result of fluctuations in the sun's activity rather than the burning of fossil fuels. For example, during record sunspot activity in the Middle Ages, there was an extended period of extremely warm global temperatures. During a recorded time of extremely low sun output activity from 1790 – 1830, Europe recorded record lows in what became known as the "little ice age". Neither of these global climate changes could have anything to do with widespread burning of fossil fuels. Furthermore, from 2002 until 2010, the evidence shows a definitive trend of global cooling, not warming, yet the carbon dioxide level is either unchanged or increasing in this same time period. Since the sun is again entering a cycle of low activity, global cooling is likely to continue for many more years. Thus the fear of "global warming" has been changed to the fear of "climate change".

**While the Earth remaineth, seedtime and harvest, and cold and heat, and summer and winter, and day and night shall not cease.**

**~ Genesis 8:22**

Did you know that the fossil record reveals that dinosaurs walked among the flowering plants such as poppies, rhododendron, and pachysandra? Carl Sagan of "Cosmos" wrongly told viewers that "the dinosaurs perished around the time of the first flower." Evolutionists believe that plants have changed dramatically over time, i.e. plants have evolved. However, paleontologists are now admitting that the fossil record shows the remains of plants very similar to modern plants all the way back to the time of the dinosaurs. Plants really show very little evidence of evolution through time.

The fossil record reveals that dinosaurs would have not only tiptoed through the flowers but wandered around among flowering trees very similar to today: dogwood, tulip trees, magnolias, walnut trees, oak trees, poplar, sweetgum and sassafras trees. Fossilized pine cones reveal that dinosaurs also lived in forests of sequoia trees and dawn redwoods. Dinosaurs lived with all the major groups of plants we find living today.



So in 200 million years of supposed Earth history (when the "age of the dinosaurs" supposedly began), why don't we find plant evolution in the fossil record? We don't find plants changing dramatically over time because God created all plants on the third day of the Creation week. When God made plants, all the different basic types of plants immediately existed. We do find different types of plants in different rock layers, and evolutionists line these up and make up a story to try and explain how one turned into another...but these are just plants buried at different levels, times, and places during the flood. The same plants we see today have been growing all around us since the beginning.

Dinosaurs really did tiptoe through the tulips!

**And the earth brought forth grass, and herb yielding seed after his kind, and the tree yielding fruit, whose seed was in itself, after his kind.**

**~ Genesis 1:12**

We all know that fingernails and toenails are there to protect our fingers and toes; they act like little tiny hardhats. But why do fingernails, when torn or nibbled, tear across the nail instead of down towards the nail bed? Scientists have found that it actually takes twice the energy to cut nails lengthwise as compared to crosswise. Nails are made of 3 layers of keratin. The top and bottom layers are found to have randomly arranged fibers of



keratin; this gives the nail its strength. The middle layer holds the key to the nail tearing crosswise; the keratin fibers run parallel to the base of the nail.

Can you imagine if our nails could easily tear lengthwise down to the nail bed? We would be in agony most of our lives from the pain. God, in His wisdom, takes care of even the smallest details of our lives: the way a fingernail tears. So, the next time you snag a fingernail, thank God that it tears across the nail instead of lengthwise.

Thou hast granted me life and favour.

~ Job 10:12

The bunchberry dogwood plant has a built-in trebuchet. A what you ask? A trebuchet is a medieval projectile-launcher. Envision it as a complex catapult. During the Middle Ages, trebuchets were designed using the principles of leverage to propel large objects faster and farther than a simple catapult. Bunchberries carpet the spruce-fir forests of North America in late April to May. The flower centers (where the pollen is produced) are only 0.1" tall so the pollen needs to be propelled upward to reach other flowers. High speed video cameras were needed to see how this plant sends its pollen flying. Scientists first tried with a camera that was able to take 1,000 pictures a second, but this proved to be too slow to capture the trebuchet's speed. So they brought in cameras capable of capturing 10,000 pictures a second. What they saw amazed them.

As bunchberry flower petals open, each petal is separated and flipped backwards (out of the way). Almost instantaneously, the stamen unfurls, catapulting pollen into the air and sending it off to pollinate other bunchberries. What is so amazing is the speed at which this process takes place. The unfurling of the petals to the launching of the pollen takes only four-tenths of a thousandth of a second. This causes the pollen to be subjected to 2,400 times the force of gravity. This is quite a projectile launcher!

Building a trebuchet took planning and design in the same way the bunchberry's miniature trebuchet took planning and design. When you see a design, you know there must be a designer. It is difficult to imagine how a plant evolved to have each of its individual petals rapidly flip down and out of the way at just the right time. If the stamen's miniature trebuchet was not ready to fire, nothing would happen. In the same way, a rapid fire pollen launcher would not be needed unless the petals burst open at the right time. Everything has to be coordinated perfectly from the beginning to do exactly what is needed.

**Great is the Lord, and most worthy of praise.**

**~Psalm 48:1**

Evidence for the Ice Age has been gathered from ice cores from Greenland and Antarctica. The Ice Age was a real event of Earth's



history, but just how long did it last, and how long ago did it occur? Evolutionary scientists assume each distinct layer from drilled ice cores represents one year. One well known core was collected in East Antarctica by the Russians and is called the Vostok Ice Core. This ice core was 6,249 feet long – more than a mile deep! The clear upper part revealed easy to see distinct layering while the lower layers were blurry. Even though the lower layers were blurry, evolutionary scientists still counted each individual layer - which resulted in thousands of years added to the assumed age of the Vostok Ice Core. The core is therefore assumed to represent 100,000 years. Yet the Bible would indicate that a worldwide flood occurred approximately 4,400 years ago, and the ice must have formed immediately following this flood.

Evolutionary scientists count the layers as years while creation scientists count the layers as snow events. Which is correct? In 1942 during World War II, six brand new P-38 Lightnings and two B-17 flying Fortress aircraft took off for a bombing mission in Germany. Unfortunately, the squadron became lost in a blizzard, and the pilots were forced to ditch the planes 18 miles from the coast of southeastern Greenland. The crew was rescued, but the planes were left. In 1980, two businessmen decided to recover the planes, thinking all they would need to do was to dust the snow off the wings. Twelve years and millions of dollars later, they finally recovered one of the planes. Why did it take so long? The planes were encased beneath 250 feet of ice in only 44 years! Obviously, the ice layers revealed snow storms and not years. The Vostok ice core does not reveal 100,000 years of Earth history but 100,000 snowstorms in a brief period of Earth history.

**Hast thou entered into the treasures of the snow? or hast thou seen the treasures of the hail?**

**- Job 38:22**

When you look up at the night sky, the constellations called "Ursa Major" (great bear), "Leo" (the lion), or "Virgo" (the virgin) do not really look anything like bears, lions, or virgins. We would not be able to identify them as anything but a random group of stars unless we learned they were called these names from star maps and history. So why is it that people in all times and all places have fairly similar names for these constellations? The best answer is that the original star map originated in one place and was then dispersed throughout the world.

An astronomer named E. Walter Maunder (1851-1928) provided convincing evidence that the constellations were invented by someone viewing them at the 40th parallel north latitude.[2] In his



analysis, the original star map could not have been produced in Egypt, Babylon, or India but must have originated farther north. Interestingly, the 40th parallel runs just north of Mount Ararat, the place where Noah's Ark landed.

Maunder further calculated that the original star map had to have been produced in about 2,500 B.C., about the time the Flood ended. This means that Noah, Shem or one of their descendants could have designed the original star map. It is even possible that the star map was based on the work of astronomers before the Flood. We do not know the original names of the constellations, but the names passed down to us include Hercules with his foot crushing the head of the Dragon and the victorious Orion holding in one hand a club and in the other the head of a lion. Later in history, Jesus (Orion) emerged victorious over Satan (Lion), who "prowls around like a roaring lion looking for someone to devour" (1Peter 5:18). Did these ancient peoples use the stars to proclaim future events that had been revealed to them by God? During the time of the tower of Babel, did their meaning become twisted from the original? These are interesting questions to ponder.

**Which maketh Arcturus, Orion, and Pleiades, and the chambers of the south. ~ Job 9:9**

Many books have been written on the mystery of how the pyramids were constructed, but the Sahara Desert has an even greater mystery. It is the mystery of mouse pyramids.

The Sahara Desert is a strange place to find mice building stone pyramids in front of their burrows, but that is exactly what they do. These mice pile up round stones, making small pyramid-like structures through which the air can flow. As the sun rises, the air heats up, but the rocks remain much cooler well into the day – having been chilled by the cold desert night. Water in the air



begins to condense on these cooler rocks, and the mice lick the water off of the rocks. This is how mice get their water! Mice in Australia also build stone pyramid-like structures to catch the morning dew. They make huge piles of round stones over their burrows; some may be up to three feet across. Again, as the sun rises and heats the air, the dew collects on the cooler stones, providing the mice with water.

How did mice "know" they could catch dew by building stone pyramids? How did two species of mice, one from the Sahara Desert and one from Australia, come up with the same idea for getting water? This is only a mystery to those who want to explain things without God (evolutionists). God knew what the mice would face and built such knowledge into their DNA, so they would be able to survive. God's purposeful programming is evident!

**He is before all things and in Him all things consist. ~ Colossians 1:17**

Due to record spring rainfall, water was diverted around the dam in Canyon Lake of Central Texas in the summer of 2002. This flood flow carved through solid bedrock resulting in a two-mile-long canyon up to 20 feet deep within a three-day period. Enormous boulders were tumbled like marbles and enough sediment to fill a thousand football stadiums wa,s almost instantaneously displaced to downstream locations.

This tiny little geological event was a drop in a swimming pool compared to the type of region-wide erosion happening all over our planet during Noah's flood. Our entire planet was resurfaced by flood flows a million times more powerful and widespread than the Canyon Lake release.

A recent scientific journal study highlighted this event as, "offering useful insights into ancient megafloods, both on Earth and on



Mars, and the deep canyons they left behind." In the words of researcher Michael Lamb, "We're trying to build models of erosion rates, so we can go to places like Mars and make quantitative reconstructions of how much water was there, how long it lasted, and how quickly it moved." [1,2]

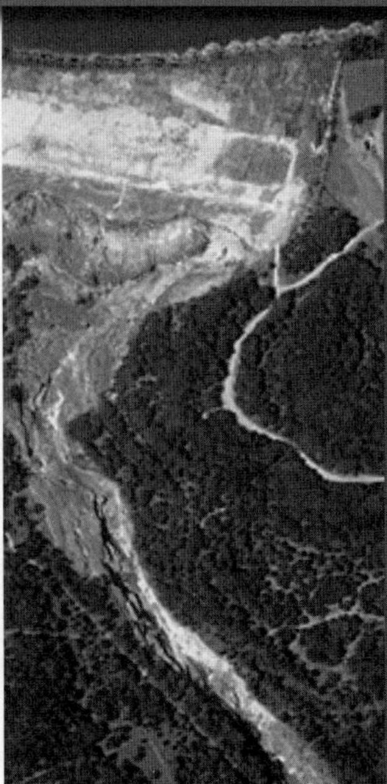

During Noah's flood, more than enough water poured off the continents to explain the Earth's sedimentary geological features, yet this explanation for sediment layers is largely ignored. Meanwhile, Mars' landscape, where there is no surface water, is attributed to exactly such an enormous, planet-encompassing flood!

**Whereby the world that then was, being overflowed with water, perished.**

**~2 Peter 3:6**

When was the last time you thought about your tonsils? They are designed to defend your body against disease. However, for over 100 years, tonsils were thought to be of no use. Throughout the 1940's – 1970's, thousands of people were put though needless pain and cost as their tonsils were routinely removed by doctors. These doctors had been wrongly taught that tonsils were useless organs "left over" from our evolutionary past.



Today, scientists have discovered that tonsils are a vital part of our immune system. Tonsils are part of a large group of lymph tissue located deep within the throat. This lymphatic system protects you from bacteria, viruses, and other invaders. When your tonsils swell, they are doing battle with germs. Scientists have found that your tonsils form a protective ring of lymphatic tissue around the opening between the nasal and oral cavities and the pharynx. This ringed boundary helps provide protection against germs in the nose and mouth. Tonsils also keep an infection local and limit the possibility of it becoming systemic by confining germs to one area of your body. Tonsils corral germs and limit the number of bacteria entering the bloodstream. Tonsils form the first line of defense for restraining the spread of viruses and bacteria. No knowledgeable doctor would today remove them because they become frequently inflamed.

The next time you have swollen tonsils, thank God for the localized disease fighting system that He has set in place for your protection.

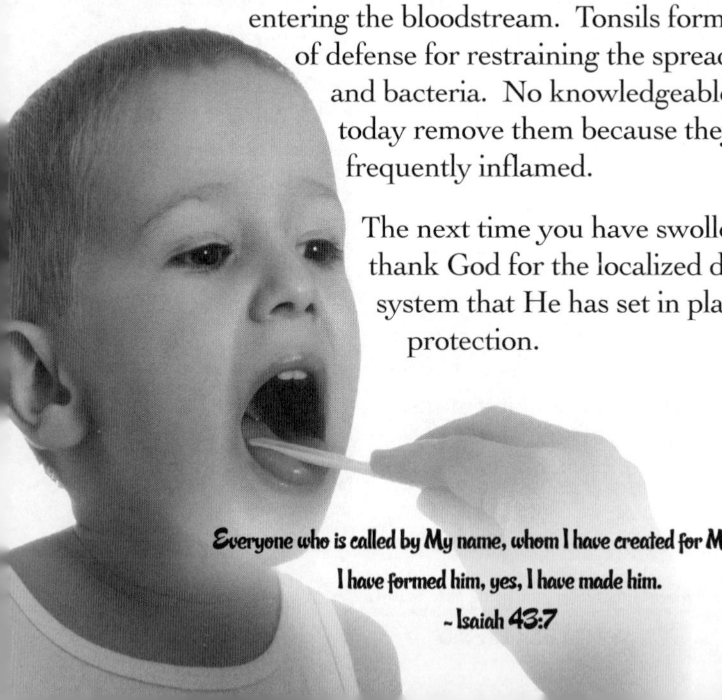

***Everyone who is called by My name, whom I have created for My glory;***
***I have formed him, yes, I have made him.***
***~ Isaiah 43:7***

Have you pondered the evolution of the Coca-Cola can? Here is how it happened:

Billions of years ago there was a rock, and from this rock oozed a cool, sweet, brown liquid. As time passed, aluminum formed around this sweet liquid forming a can. Millions of years later, red and white paint fell from the sky forming the words, "Coca-Cola, 12 fluid ounces."

Is this true? Of course not! You know when you see a Coca-Cola can that there must be a Coca-Cola maker. It did not happen by accident or chance.

In the same way, when we examine the eye with its many parts, such as the cornea, iris, lens, retina, optic nerve, and the right amount of fluid pressure in the vitreous humor, we know that this could not happen by accident or chance over millions of years. Our eye needed all its parts right from the beginning in order for the eye to see. It is irreducibly complex. If we do not have the right amount of fluid pressure in the vitreous humor, we would experience glaucoma and blindness. If our lens were cloudy, we would experience cataracts and the inability to see. Each part of the eye had to be in its proper place and fully functioning right from the beginning in order for us to see. A seeing eye could not happen by blind chance.

Without our eyes, we could not even observe a can of Coke to know that it was designed. Yet, the evidence that the eye is



designed is even more overwhelming than the Coke can. When we see an eye, we know there must be an eye-Maker. That eye-Maker is God.

**He that planted the ear, shall he not hear? he that formed the eye, shall he not see? The Lord knoweth the thoughts of man.**

**~ Psalm 94: 9b,11**

As the American alligator glides toward its prey, he makes no sound, only a few ripples on the water's surface. The gator appears to be a floating log, unassuming…until it lunges with agility and speed to catch his prey.

This predator is designed for survival in swampy waters. His eyes, nose, and ear slits are on the top of his head, allowing most of his body to remain submerged. When he plunges into the water, the nose and ear slits are designed to close, keeping the water out. To find his prey underwater, an alligator has built-in goggles:



clear nictitating membranes that cover his eyes. To aid in hunting, gators have sensory pits along their jaws that allow them to pick up the slightest disturbance in the water. When he drags his prey under the water, why doesn't the water flood into the gator's throat causing him to choke? Because he has been designed with a special valve at the back of his throat that allows him to open his mouth and eat his food underwater without choking!

Once the prey is eaten, the alligator needs to digest it. Digestion requires the input of heat energy, and alligators rely on heat from the sun to digest their food. Yet, alligators are cold-blooded and often live in relatively cool waters. Therefore, alligators have built in solar panels on their backs: bumpy scales full of blood vessels that catch and draw in heat quickly. This heat helps them stay warm and digest their food.

What if the alligator did not have these solar panels on his back, how would he digest food? What if the alligator did not have his underwater goggles, how would he see to get his prey? What if he did not have the check valve to stop water from going down his throat while eating his prey underwater? These features all show design. The alligator is designed by God for his life in a watery swamp.

**Hearken unto this, O Job: stand still, and consider the wondrous works of God.**
**- Job 37:14**

Empirical sciences are those branches of science which can be described and tested by mathematical equations. They make predictions that can be easily tested. For example, there are mathematical equations which can predict the exact effect of



gravity or compute how fast soft tissues from dead animals will deteriorate.

Paleontology, evolution, and creation are not empirical sciences. They are scientific disciplines which make observations from the physical world and place these observations into a philosophical framework in order to understand the significance of these observations. Although evolution has been equated with the laws of gravity and commonly referred to as "a fact" and "the foundation of biology", these claims are intellectually dishonest.[1]

What mathematical formula shows the speed at which one animal is turning into a completely different type? What equation shows the type of evolutionary change that is about to happen? What prediction of evolution explains where evolution is heading? On the other hand, one can use equations of empirical science to show that soft tissue from dead animals could not possibly survive for millions of years. Evolution requires enormous time period; therefore, no dinosaur fossils should contain soft tissue. The biblical creation model uses this same empirical science to predict that dinosaurs and other fossils buried during the flood could be found with such soft tissue still intact because they were fossilized quite recently during the worldwide flood.

Which is true? The Biblical model! Many examples of unfossilized tissue in dinosaurs have now been documented in scientific literature.[1] It would seem that creation is actually much more "scientific" than evolution.

**O Timothy, keep that which is committed to thy trust, avoiding profane and vain babblings, and oppositions of science falsely so called: Which some professing have erred concerning the faith.**

**-1 Timothy 6:20, 21**

Nature abounds with examples of natural enemies that co-exist in mutually beneficial relationships. Called symbiotic relationships (or mutualism), such pairings are a mystery to those who reject creation.

An example is the narrow-mouthed toad and the tarantula. Toads are normally a food source for tarantulas, but the narrow-mouthed toad lives right inside of the moist tarantula burrow where it eats ants and other pests that typically feed on tarantula eggs. Thus, the spider trades an easy meal for pest protection, and the toad places



itself in mortal danger with a natural enemy in exchange for free room and board.

A recent evolution-based journal claims that evolution explains how such relationships develop in order to benefit both species. Yet, the research explained nothing. It was noted that when other similar sized frogs and toads were placed in the spider nest, either the frogs or the spider immediately moved out. Many hypothetical examples were given of how chance occurrences could benefit two normally hostile species, but this storytelling explains nothing about how such "learned" benefits could be passed onto the next generation. Furthermore, the article admits that experiments simulating these chance occurrences consistently fail to result in any symbiotic relationship. Thus the benefit (i.e. creation or design) of such relationships is clearly seen, but any natural origin simply remains a mystery.

The mystery is immediately solved once a creator of the symbiotic relationship is acknowledged. Once again we see the fingerprint of God.

**The fear of the Lord is the beginning of wisdom:**
**and the knowledge of the holy is understanding.**
**- Proverbs 9:10**

One of the objections which evolutionary geologists frequently make to the Biblical interpretation of world history is the lack of time needed to form the geological features of our planet. One example is Yosemite Valley. This valley is acknowledged to have been completely filled with snow, compressed into a solid glacier of ice over 3000 feet deep. That would require approximately 50,000 feet of snowfall (adding extra snow to account for meltback). If the Earth is only 6000 years old, how could so much snow have accumulated so rapidly?

Larry Vardiman recently modeled this possibility using conditions expected after the worldwide flood to predict the snowfall in the Yosemite region. Massive continental land movements and volcanism would have left the ocean significantly warmer after the flood. Warmer oceans result in vastly greater evaporation that would fall as gargantuan snowstorms at higher elevations.



Even today, weather patterns known as Pineapple Express storms can dump as much as 20 feet of snow in the higher Sierra Nevada Mountains in one week! If the ocean surface temperature was elevated to 113°F (this is the type of ocean temperature possible by the end of the worldwide flood), these same weather patterns would increase in both frequency and output with models showing each storm dropping as much as 80 feet of snow per week. Thus, 50,000 feet of snow could easily accumulate from 625 snowstorms. If there was an average of 10 such storms per year – the Yosemite glacier could have formed in fewer than 60 years. These snowstorms would have continued until the ocean cooled. The Yosemite glacier likely took much longer than 60 years to form but not thousands of years. The Biblical model clearly explains the rapid formation of the Yosemite geological formations – it was a direct consequence of the ice age which immediately followed the flood.

**Out of whose womb came the ice? and the hoary frost of heaven, who hath gendered it? The waters are hid as with a stone, and the face of the deep is frozen. - Job 38:29,30**

There are only two possibilities for explaining the origin of everything:

| Evolutionary story: | Biblical account: |
|---|---|
| 1. Sun before Earth | 1. Earth (day 1) before sun, moon, stars (day 4) |
| 2. Dinosaurs evolved from birds | 2. Birds (day 5) before dinosaurs (day 6) |
| 3. Earth initially hot & molten | 3. Earth, a water surface (day 2) then land (day 3) |
| 4. Flowering plants evolved 220 million years after other plants | 4. Flowering plants with other plants (day 3) |
| 5. Sun before plant life | 5. Sun (day 4) after plant life (day 3) |
| 6. Fish before fruit trees & birds | 6. Fish & birds (day 5) after fruit trees (day 3) |
| 7. Man evolved from lower animals | 7. Formed from dust and has the "breath of life" |



This is just a partial list of the obvious contradictions between evolution and creation. The biblical order of events simply does not match up with evolution. Any combination of evolution with God's Word destroys the credibility of God's Word. Compromising with truth is the same as destroying the truth. As Bible-believing Christians, we must take God at his word.

**For in six days the LORD made heaven and earth, the sea, and all that in them is, and rested the seventh day: wherefore the LORD blessed the sabbath day, and hallowed it. ~ Exodus 20:11**

Have you ever looked closely at a tick? The male wood tick has "suspenders": two parallel, white lines on its back running from "head to tail". The female wears a white "necklace" or has a white spot, depending upon the species.

If you live in the country, you've probably had your fill of these little bloodsuckers. Yet, ticks are master chemists. The tick's saliva produces an anesthetic which allows the tick to attach itself to our skin and sip our blood without us noticing it. Unlike a mosquito (which can often be felt), or a bee sting (which is definitely felt!), a

## INSPIRED BY BIOLOGY

AUGUST 16TH

tick can attached itself without being noticed. Once attached, the tick performs another marvel of chemistry – it turns off our blood's clotting mechanism. Normally, our blood clots when a wound happens; however, the tick's saliva is able to disable this clotting mechanism, thereby having a continual flow of its delicious meal.

Could random natural processes have produced these complex chemical reactions? Not a chance. Our Master Designer created this amazing saliva. It may have had some alternate use before the Fall, but it is still a marvel of chemistry even in our world. Maybe the next time you find a tick sucking your blood, it will help to consider it a marvel of creation and chemistry rather than an annoying little pest.

But probably not.

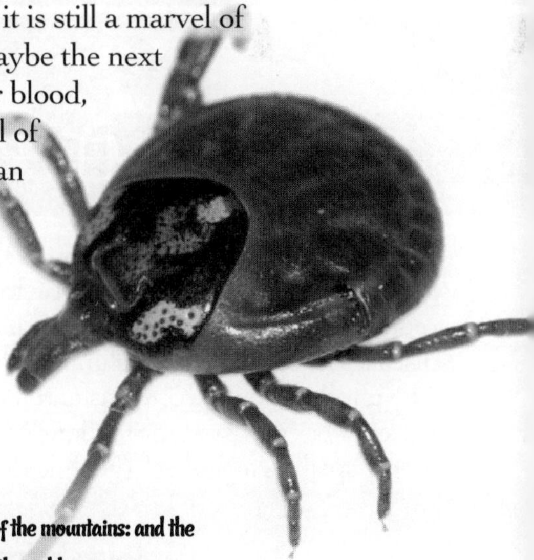

**I know all the fowls of the mountains: and the wild beasts of the field are mine.**

**~ Psalm 50:11**

There are currently 4,000 known genetic diseases which have accumulated on the DNA code of human beings – all caused by single gene defects.[1,2] All of these mistakes are harmful (frequently fatal) to those born with these permanent errors to their DNA. Those manifesting genetic diseases have misinformation within the DNA of every cell of their body.

We are all carrying around part of this damaged information, like a book containing thousands of misspelled words and grammatical errors. All it would take for the next generation to be born with one of these genetic errors is for the father and mother of the child to have the same genetic mistake on the corresponding copy of their "book" (genome). This is why close relatives (people carrying the same spelling errors in their closely related books) cannot get married and have children. The result is an inevitable myriad of health problems, mental illnesses, or genetic defects within their children. If every new generation has an increase in the number of genetic mistakes, there must have been fewer mistakes in the past. The farther back we go, the more perfect our DNA code must have been. New genetic diseases, unknown to previous generations, have developed repeatedly within recorded medical history. However, genetic diseases have never been eliminated from the human genome. The implications are obvious and inescapable – humans could not possibly have existed for millions of years, or there would be far more "spelling mistakes" (genetic diseases).

AUGUST 17TH

# MICROBIOLOGY INSPIRED BY

**But this I say, brethren, the time is short:**

**...for the fashion of this world passeth away.**

**-1 Corinthians 7:29, 31**

Perspective matters. Many compromising Christians claim it does not matter whether we believe God created recently or just started everything rolling billions of years ago. This parable illustrates why the truth matters.

## INSPIRED BY CHRISTIAN TRUTH

AUGUST 18TH

Years ago two boys lived with families which were barely eking out a living. Both boys longed to own their own bicycle but knew that there was no possibility that they could ever afford one. One day the grandfather of the first boy came to visit. They spent time together and got to know one another. He listened as his grandson talked about his life, his adventures, and his desires. He took time to play with his grandson, interact with his grandson, tell him stories, and impart wisdom. After the long visit, the grandson tearfully said goodbye to his grandfather as the older man climbed into his car to drive home. Yet, the tears really flowed as the young boy went into his room and found a brand new bicycle.

The second boy came from a home steeped in family heritage. Night after night at the dinner table, he was told stories about his rich grandfather. One day the boy noticed something sticking out of the ground near the ruins of his grandfather's shed. His curiosity got the better of him, and he dug the object from the ground. To his amazement, it turned out to be the handlebars from a bicycle. As he continued to dig he found a pedal, a wheel, a frame, tires, brakes, grips… everything he needed to construct his own bicycle. For weeks the boy spent every waking moment working on his bicycle. He polished, repaired, fixed, and patched. Finally he succeeded in assembling all of the parts left by his grandfather into a refurbished, working bicycle.

Both boys ultimately received a bicycle from their grandfathers – one as a recent and direct gift, the other from an assembly of parts from a distant and unknown stranger. Which child would have the greatest devotion and desire to please his grandfather? Adding enormous time and distance to any relationship cannot help but diminish its relevance.

**Though the LORD be high, yet hath he respect unto the lowly: but the proud he knoweth afar off. ~ Psalm 138:6**

Dams serve important functions in human cultures. They provide lakes (for drinking, irrigation, and recreation), they prevent floods by slowing the overflow of water onto downstream flatlands, and they generate electricity through the force of the flowing water through the dams. Yet mankind is not the first to use this ingenious engineered structure. Long before mankind was building dams, the beaver was designing and building dams. But have you considered the design of the beaver itself?

- The beaver has special valves in his ears and nose.
- He has built in goggles for seeing clearly underwater.



- He has specially designed self-sharpening front teeth.
- During the winter, he retrieves his stored food underwater. When he opens his mouth, does water flood in? Two specially designed folds of skin, each on either side of the mouth, meet behind the front teeth and seal off the rest of the mouth.
- The beaver's large paddle-shaped tail is used as a rudder. When he is swimming with a branch in his mouth, the tail compensates for any uneven drag from the branch.
- The rear feet are large and webbed - like duck feet – giving the beaver the force needed to transport large loads.
- The beaver is kept waterproof by oil from two oil glands and its two layers of fur.
- Large lungs and liver allow for storage of more air and oxygenated blood while he's underwater.

All of the beaver's features must have been present and fully functional from the beginning for it to survive its semi-aquatic life.

*Declare his glory among the heathen, his wonders among all people.*

*~Psalm 96:3*

The salt used on your fries probably came from an enormous deposit covering an estimated 600,000 square miles deep below northern Ohio and New York. This salt is so pure that it can be ground up and immediately used for human consumption – containing essentially no contamination from sand, wind-blown dust, plant or animal fragments & debris, etc…

How did such an enormous deposit form? The standard explanation is that a salt water depression repeatedly filled and evaporated over huge periods of time, leaving behind this enormous salt deposit (called an evaporite formation). Yet the ultra-pure characteristics and massive size of this deposit stretch the credibility of such a scenario. One ancient deposit called Zechstein of Europe exceeds 6,500 feet in thickness. To produce this evaporite deposit by evaporation of seawater would have required a column of seawater 80 miles deep (The deepest ocean basin is only 7 miles)! Evaporation



of all the water in the Mediterranean Sea, would only yield a deposit 196 feet thick.

The Bible provides a better explanation for such deposits. During the worldwide flood, enormous vents opened up deep within the ocean as the continents were sliding to new positions. Even today ocean vents spew forth supersaturated, mineral-laden fluids – but on a much smaller scale than what was happening during the flood. It is possible that an extremely supersaturated solution of salt formed in localized areas during the flood, and as the supersaturated solution cooled, ultra-pure salt precipitated to form widespread salt deposits. These salt concentrations were then covered by flood sediment. These thick beds of salt would have formed in a short time in the worldwide flood of Noah's day. This explains the thicknesses, purity and worldwide nature of these deposits. As you reach for that salt shaker, realize that that salt was probably mined from a deposit laid down during Noah's Flood.

**In the six hundredth year of Noah's life, in the second month, the seventeenth day of the month, the same day were all the fountains of the great deep broken up, and the windows of heaven were opened.**

**~ Genesis 7:11**

Dinosaurs provide a great test for which view of human history is correct.

- If the evolutionary view is correct then dinosaurs died 65 million years before mankind appeared, and there should be little or no knowledge of such creatures.
- If the Biblical record is correct, then they were created along with mankind, went through the flood with Noah, and have only gone extinct within the last few thousand years. Therefore, there should be extensive evidence of mankind's knowledge of dinosaurs.

The biblical model wins hands down. The word 'dinosaur' was not coined until the 1800's, but cultures around the world have historical references to the existence of great beasts known as 'dragons'. There are ancient drawings, rock and cave etchings, eyewitness accounts, historical records, old literature, pottery, burial artifacts and folk stories that exactly describe the appearance of various dinosaurs. Some of these artifacts even provide knowledge about the appearance and behavior of these creations in advance of scientific discoveries from their fossil impressions. Even the Bible contains a totally accurate description of a sauropod-like dinosaur, which includes its appearance, size, and habitat. It is the widespread evidence of such knowledge that proves the biblical model to be accurate.

**Behold now behemoth, which I made with thee; he eateth grass as an ox.**
**16Lo now, his strength is in his loins, and his force is in the navel of his belly...He moveth his tail like a cedar...His bones are as strong pieces of brass; his bones are like bars of iron...He lieth under the shady trees, in the covert of the reed, and fens...the willows of the brook compass him about...**
**Behold, he drinketh up a river, and hasteth not: he trusteth that he can draw up Jordan into his mouth.**

**~ Job 40:15-23**

From wisdom teeth to our appendix, from tonsils to our tail bone, almost all the features in the human body that have been historically promoted as "non-functional" are now known to have a useful function. In 1893 there were 83 human features considered to be useless leftovers from our ancestral past; it has since been acknowledged that all 83 have legitimate, useful functions.

The latest feature to be removed from the vestigial "useless feature" list is the tiny hair-like structure sticking out of the cells in the human kidney, the primary cilium. Since these hairs looked similar to the cilia used for movement of single-celled organisms, the cilium was labeled as a useless leftover from our evolutionary past. For years, these structures were promoted in biology books as evidence for evolution.

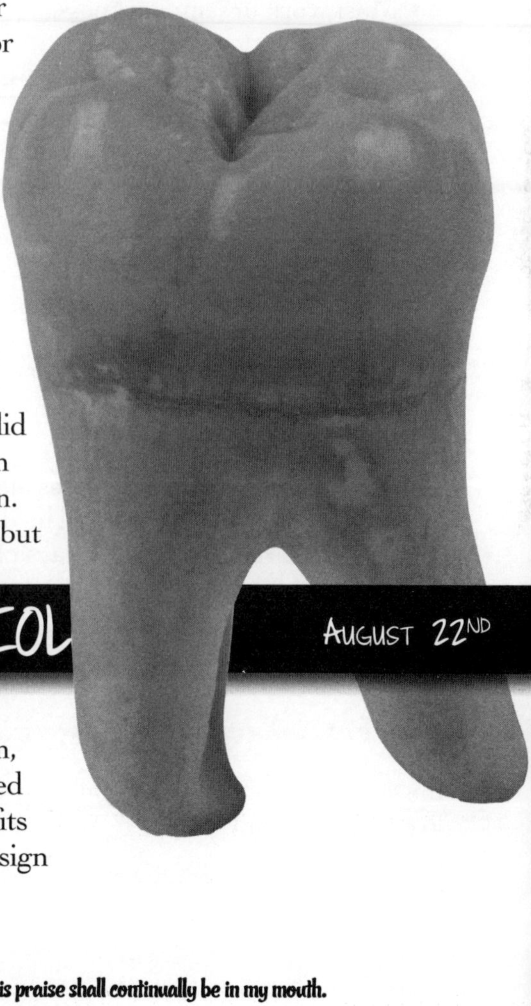

In 2010, it was discovered that the hair-like structure was not a throw-back to our evolutionary past but a signal receiver allowing cells to communicate and interact with other cells in the body.[1] If we did not have cilium, communication between cells would not happen. These structures are not usless but



very much needed! Once again, the belief in evolution has slowed discovery; the actual evidence fits perfectly with the intelligent design of life.

**I will bless the Lord at all times: his praise shall continually be in my mouth.**

**-Psalm 34:1**

Pretend you are responsible for producing a miniaturized navigational system for a tiny spy drone plane with the following requirements:

Lens:

- Must be able to see in all directions at the same time.
- Must be able to see all the colors of the rainbow and ultraviolet light.
- Must be able to perceive objects as small as 0.04 (the width of a hair) inches from 20 feet away.
- Must convert images into electronic signals and send them to a central computer for instant processing decisions.
- Must work in semi-darkness.

Navigation system:

- Must detect the direction of the Earth's magnetic field.
- Must detect and navigate changes based on the sun's current position.
- Must be capable of navigating to a new location as far away as 3,000 miles.

Other requirements:

- The entire system must be designed to weigh less than 0.5 grams.
- The system must be smaller than a pea.
- The system must be built in fewer than 7 days by one person in total darkness.

This optical/navigation system has already been designed and manufactured in massive quantities. The design is found in a monarch butterfly. During the first six days of creation, God was also effortlessly designing a billion other creatures, overseeing every detail of



the cosmos, while foreseeing all future events (including each of our lives). Still want God's job? Maybe you'd better just stick with the one He has given you – obeying Him!

**According as he hath chosen us in him before the foundation of the world, that we should be holy and without blame before him in love. - *Ephesians* 1:4**

Did you know that the African gazelle has a radiator in its nose? When chased by a lion, the African gazelle can reach speeds up to 50 mph. This raises the gazelle's body temperature. However, when scientists measured the temperature of the gazelle's brain after extreme exertion, they found the temperature to be remarkably low. Scientists were surprised to find that the gazelle has its own totally unique blood cooling system.

Before the gazelle's blood goes to the brain, it passes through a cool pool of blood near the nasal cavity. This allows the gazelle to be



"cool-headed" in the heat of the flight from a lion. How long did it take a gazelle to come up with this radiator in its nose? Picture this: Mr. Gazelle is being chased by a lion; without the radiator, he boils his brain and dies. If he slows down (in order to not overheat), he is caught by the lion and dies. Either way, unless everything is in place, he dies.

What Mr. Gazelle needed was a radiator allowing him to escape predatory animals. And that is exactly what the Master Designer gave him. It was preprogrammed into the gazelle at the time of creation and ready for future environments after the Fall when he would become a meal for lions. God thought of everything in advance.

**Remember his marvelous works that he hath done; his wonders, and the judgments of his mouth. - Psalm 105:5**

Deep within the mantle of the Earth, 1800 miles below the Earth's surface, melted rock is estimated to be 5000°F. However, scientists recently found a continent-sized block of rock within this inferno which is 3000°F cooler than the surrounding material. If the Earth's interior were billions of years old, there would not be a large area with a completely different temperature than the surrounding rock. The heat would have been uniformly distributed over time.

According to the theory of plate tectonics, the Earth's continents are thin overlapping plates floating on thicker, denser material. At the overlap of the individual plates, the lower plate is plunged deep into the Earth. New ocean floor is forming, slowly shoving these plates forward. If this process had taken place over millions of years, the subducted rock plates would have had plenty of time



to heat up to the surrounding temperature deep within the Earth. Yet, an enormous block of cold material is located deep within the Earth at this subduction location – as if an entire continent had been rapidly and recently plunged into the Earth. The Bible provides a solution for this mystery.

It has been demonstrated that there could have been rapid continental movement during the flood of Noah only 4500 years ago. During this world-wide catastrophe, the North American continent could have separated from Europe/Africa and moved into its current position in a matter of months while plunging the pre-flood rock layers deep within the Earth. This explains how a large, cooler area of rock could exist deep within the Earth. But, this also means the Earth could not possibly be millions of years old.

**...the world that then was, being overflowed with water, perished.**

**- 2 Peter 3:6**

Have you considered why God made other planets besides earth? If other planets do not support life, why do they exist? First, the planets help confirm the passage of time. The moon and sun provide the primary chronology for time passage, but other planets provide a confirmation. Another purpose for the planets is that they act to stabilize the solar system. This fact was not discovered until the 1790s by Pierre-Simon Laplace. Each planet was created by God with a specific size, gravity, rotation rate, magnetism, chemical makeup and axis tilt. Each specific property helps keep the solar system safe and stable. The sun's gravity pulls on the planets while the planets pull on each other.

Gravity is the mysterious force that acts through great distances of empty space. The gravitational attraction between the Earth, the sun, and the other planets is continually adjusted as their locations change in their orbits. Their elliptical orbits prevent them from plummeting into each other, but it is gravity which is the "glue" that holds our solar system and universe together. The other planets in our solar system exert a gravitational pull that helps stabilize the Earth's distance from the sun. This keeps the Earth from approaching too close or moving too far from the sun. Planets were created by God to keep Earth in both a stable and safe orbit.

**Dominion and awe belong to God; he establishes order in the heights of heaven. ~ Job 25:2**

When viewed from outer space, our spherical Earth appears like a disk or a circle floating in emptiness. The Bible refers to the "circle of the Earth", which is a perfectly accurate description of the appearance of the Earth from outer space. Job 26:7 states that God "hangs the Earth on nothing." In your mind, you can envision an Earth floating in space, the Earth literally hanging upon nothing. The Earth can hang upon nothing because it is held in place by the mysterious force of gravity. Sir Isaac Newton identified this force as gravity in 1650, some 3,500 years after it was described in Job.

People have developed many myths concerning the structure supporting the Earth. For example, the Hindus believed the Earth was supported on the backs of four elephants. The Vedic priests had the Earth set on 12 solid pillars. The peoples of Northern Siberia had the Earth supported by 3 great fish. The many tribes of Eurasia believed the Earth was supported by a great bull. Throughout history, people have viewed the Earth as sitting upon something – only God's Word spoke the truth. It was not until the 1960s, when we went out in space and looked back at Earth, that we saw how "the Earth hangs upon nothing", and we viewed the "circle of the Earth". Whenever God speaks on science…it is true!



# BIBLICAL ACCURACY

INSPIRED BY

It is He who sits above the circle of the earth… ~ Isaiah 40:22

A mile below the ocean's surface, amidst crushing water pressure and total darkness, lives the Angler fish. On the forehead of the female Angler is a "fishing rod" with an "artificial worm". She dangles this bait over her mouth to catch her dinner, but how can her prey find the bait in the deep darkness? The dangling "worm" comes with its own light! This light displays sophisticated technology; it gives off no heat. This heatless light is produced when a compound called luciferin is oxidized with an enzyme called luciferase. How could a deep sea fish evolve an "artificial worm" with a high-tech light, which is dangled over its mouth?

## INSPIRED BY BIOLOGY

AUGUST 28TH

How could it have learned to manufacture and separate two necessary chemicals – either useless without the other? All these features had to be present from the beginning in order for the Angler to catch its dinner.

The Angler fish also has an interesting way of reproducing. In the deep-sea darkness, it would be difficult for the males and females to find each other. The males do not have a digestive system but do possess the ability to smell pheromones released by the females. When it finds a female, the male bites her on the abdomen or flank. Once his jaws lock shut, they never let go. Soon an enzyme from the male digests the skin of his mouth, and the two become one. At this point, the pair drops to the bottom of the ocean.

Why does the female allow the male to bite her? What evolutionary mechanism enabled the male's circulatory system to merge with the female's? Evolution does not have an explanation. God designed this Angler fish to show His creativity.

**...light is come into the world... ~ John 3:19**

In order for humans to gracefully move around on two legs, they must have a greater sense of balance than those creatures moving on four legs. When a dog or cat walks on all fours, the center of balance is within the four points of contact on the ground. Contrast this with the center of balance on two legs; this center of balance is between the two feet and represents a relatively smaller area. Humans achieve this balance primarily through inner ears.



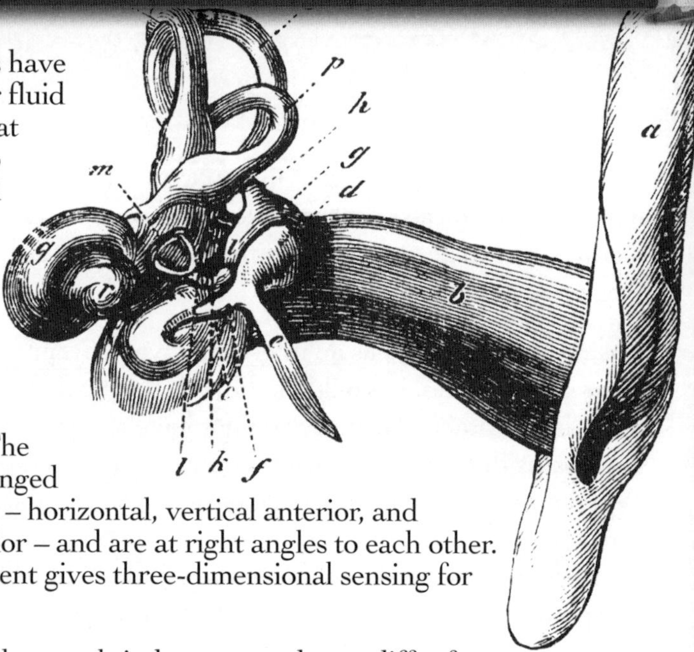


Our inner ears have 3 semi-circular fluid filled canals that are sensitive to movement and gravity. Fine hairs in the canals send out signals to indicate the direction and speed of head movements. The canals are arranged in three planes – horizontal, vertical anterior, and vertical posterior – and are at right angles to each other. This arrangement gives three-dimensional sensing for balance.

The semi-circular canals in humans and apes differ from each other. Humans have two large vertical planes and one small horizontal plane, while apes have three small canals of similar size. The three ape semi-circular canals are of similar size, allowing apes to be successful at climbing trees. In contrast, humans are designed for walking upright and need balance sensors specifically designed for walking upright. Thus, they have larger canals in the vertical planes. When the fossil record is examined, no transitional fossils of semi-circular canals have ever been found. In conclusion, man was made to walk upright, and apes were made to walk on all fours. This is just one a miriad of differences between the two; differences seldom shown to students.

**And he said unto me, Son of man, stand upon thy feet...**

**~ Ezekiel 2:1**

The flight of a dragonfly is a wonder of physics. Most insects use 2 wings to fly. However, the dragonfly has 4 wings that can hover or move both forward or in reverse, either quickly or slowly. Dragonflies have an unusual muscular system that allows the four wings to move independently; this is the key factor for its incredible acrobatics. Physicists are still trying to understand the forces involved, and bioengineers have been trying to build robotic versions of dragonflies. At first they assumed that the out-of-phase flapping of a dragonfly's wings came with an extra energy usage "cost". Upon further study, it was discovered that the hind wings actually used force and momentum from the wake of air sent by the front wings. This reduced the aerodynamic power requirements by up to 22% as compared with a single pair of wings. What is even more amazing is the ability of dragonflies to switch between out-of-phase flapping and in-phase flapping as needed. Dragonflies were designed to do what they do, and what they do, they do very, very well.

## INSPIRED BY PHYSICS

AUGUST 30TH

Most insects' wings are attached to plates of the chitonous exoskeleton that are, in turn, attached to muscles that move the plates that move the wings. Dragonfly wings, on the other hand, are directly connected to large muscles within the thorax. The interior of the thoracic exoskeleton is massively braced and strengthened to withstand the pressures of these large flight muscles.[2] Before the flood, some dragonflies had wing spans of 3 feet. What a spectacular sight and reminder of the Creator they must have been.

*O sing unto the LORD a new song; for he hath done marvelous things...*

*~Psalm 98:1a*

The Namib beetle lives in the dry, desolate desert of Namibia. It rarely rains there. So how does the beetle get the water it needs to survive in this desert environment?



Early in the morning, a fog settles over this desert. Most of the water evaporates with the sun and daily winds. God designed this beetle with tiny bumps on its back. The tops of these bumps attract water. Between the bumps are valleys coated with a waxy material that repels water. As the fog settles in, droplets collect onto the bumps, and when enough water is collected, it rolls down into the waxy valleys. Guess where these waxy valleys end up? In the beetle's mouth! How did this evolve? How do the bumps get coated to attract water and the valleys get coated to repel water? How do all the valleys end up at the beetle's mouth? In order to explain everything without God, evolutionism teaches that all this evolved over millions of years. In the dry and thirsty desert, all you would have would be dead beetles, not slowly evolving beetles. God's awesome creativity extends from the colossal cosmos to the bumps on a beetle.

... the Lord spake unto Moses,
"Gather the people together, and I will give them water."
~ Numbers 21:16

# SEPTEMBER

*I know God will not give me anything I can't handle. I just wish that He didn't trust me so much."*

**- Mother Teresa** (1910-1997)
Missionary to India

*God is faithful, who will not suffer you to be tempted above that ye are able; but will with the temptation also make a way to escape, that ye may be able to bear it.*

**- 1 Corinthians 10:13**

It is inherent in human nature to view all things through the lens of what we believe to be reality or truth. The natural response to anything which falls outside of what we believe to be true is the rejection of the challenging information (at times subconsciously



and instantly). This is especially true of things that threaten the very basis of what we believe to be reality.

This is the reason that very few people have any interest in attending a Flat Earth Society meeting. The concept is too far outside what we believe to be true. It is also why there is such a vehement battle over what is taught within our school system. ANY mention or evidence that we may have an "Intelligent Designer" (or even allowing the flaws of the theory of evolution to be exposed) is opposed with a vengeance. Furthermore, the courts have been used for the last 50 years to keep students from considering any possibility of creation. Thus, we now live in a society where the evidence for creation is equated with the evidence for a flat Earth.

This is the primary reason the evidence for biblical creation is so hard to consider for most people. They have been systematically trained through media and education that all things can be explained by natural causes over long periods of time. It is the filter though which people are trained to evaluate all incoming data. Evidence for the supernatural creation of life, the entrance of disease and death into creation because of the actions of mankind, a world-wide restructuring flood, and the recent dispersion of people groups across the globe after this flood is simply ignored by the majority of scientists whose worldview prevents them from even considering such possibilities. To break this cycle, scientifically solid, verifiable data must be presented in a credible, winsome, plausible, and non-insulting manner with those who have been trained to filter data using only a naturalistic mindset. Sharing this book with others can accomplish this goal.

**Thy word is true from the beginning: and every one of thy righteous judgments endureth for ever.**

**- Psalm 119:160**

In the 1600s, Dr. Jan Baptist von Helmont left wheat in an open jar with dirty underwear. Several weeks later mice appeared.



From this experiment, Dr. von Helmont concluded that mice spontaneously arose from wheat and dirty underwear. Other scientists of the day offered a second "proof" that life could arise from dead things. These early scientists left uncovered, rotting meat lying on a table. After two weeks, living maggots appeared within the meat. For many years, this was the reigning paradigm of science – life produces itself spontaneously from non-living matter.

In 1668, Dr. Francisco Redi slightly changed the maggot experiment by covering the rotting meat with cheesecloth. The result - no maggots appeared. He realized that for maggots to appear there needed to be parent flies. Finally, Dr. Louis Pasteur began a series of experiments with broth that proved once and for all that life comes from living things and dead things never come to life. Through careful experimentation, Louis Pasteur discovered the law of biogenesis (bio: "life", genesis: "beginning") that life begets life. This law of science has never been violated by any experiment or observation. Pasteur went on to apply this discovery to the process of pasteurizing milk – which has prevented millions of deaths by stopping microorganisms from growing in milk.

Ironically, the same year that Pasteur, a strong Christian believer, proved that life could never have arisen by itself (therefore life must have been made by God), Charles Darwin (angry with God over the death of his young daughter) published the *Origin of the Species* – supposedly showing how one form of life could transform itself into another. This rapidly led to the belief that life did come from non-life. That year was 1859 – a landmark year for both truth and deception.

**Am I therefore become your enemy, because I tell you the truth?**

**- Galatians 4:16**

Darwin cited the large, brilliantly multi-colored beak of the toucan as an example of evolution – produced by natural selection in order to better attract a mate. Others have suggested that the beak was produced by the random forces of nature because it allows the toucan to have an advantage when peeling fruit. New research is now uncovering that the main function of the toucan's beak is to help the toucan stay cool in the tropical heat. The toucan can flush large amounts of heat away from its body using its massive beak. The beak, which makes up 30-50% of the surface area of the bird's body, has a network of blood vessels close to the



surface. Depending on how much heat needs to be dissipated, the toucan can regulate the amount of heat radiating from its bill. Using infrared photography, researchers have observed the toucan adjusting this blood flow to its beak.

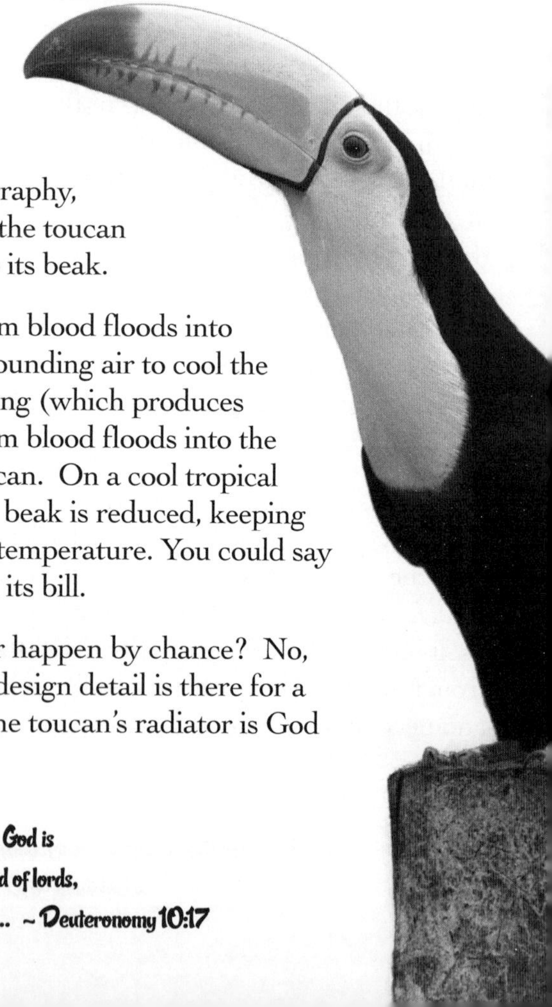

On a hot, tropical day, warm blood floods into the beak, allowing the surrounding air to cool the toucan. If the toucan is flying (which produces body heat), even more warm blood floods into the beak, again cooling the toucan. On a cool tropical night, the blood flow to the beak is reduced, keeping the bird's body at the right temperature. You could say the toucan has a radiator in its bill.

Did the radiator in your car happen by chance? No, it was designed, and every design detail is there for a purpose. The designer of the toucan's radiator is God Himself.

**For the Lord your God is
God of gods, and Lord of lords,
a great God, a mighty, and a terrible... ~ Deuteronomy 10:17**

If you listen closely, you will realize that birds are extremely skillful musicians. When birds sing, you can hear melody, harmony, and rhythm. Many birds sing in "absolute pitch", that is, if they sing a song in G major today, the same song will be sung in EXACTLY the G major pitch all other days. There is also evidence that birds can transpose songs from one key to another. Male blackbirds have been observed to sing in matched counter-singing, i.e. each blackbird takes turns singing a tune back to other birds. Some birds even sing a variation on a theme. When human musicians attempt a matching counter-singing, it requires planning, talent, practice, and design.

Some birds sing duets; the African robin performs its duet antiphonally (where 2 birds sing alternate notes in a song). To execute this type of duet requires split second timing! When you hear a bird singing, notice that the songs contain consonant intervals, which produce a pleasant sound, rather than dissonant intervals. Hermit thrushes have been found to sing the pentatonic scales. Other birds can sing two notes at the same time because of the position of their voice boxes.

Both Mozart and Beethoven recognized birds' musical abilities and borrowed bird songs as inspiration for their music. Bird songs show evidence of structure and beauty. If evolution was true and songs evolved by accident or chance, then at least half the songs should be grating, irritating, and nonsensical. We do not hear that! The worldwide morning song of birds is known as the "Dawn Chorus" - welcoming in a new day. When we hear beautiful music, we know it has been composed and that there must be a composer.

## INSPIRED BY DESIGN — SEPTEMBER 4TH

Bird songs are beautiful – take time to listen to what Jesus, the Greatest Composer of all, has composed for our pleasure.

**Let every thing that hath breath praise the LORD.**

**~ Psalm 150:6**

The brain needs oxygen, nutrients, and food, all which come from the blood supply. Thus, the brain needs a lot of blood. However, there is a problem. The brain's nerve cells (neurons) are sensitive to many chemicals contained within the blood, and direct exposure to the body's bloodstream has the potential to cause neurological problems. How would you solve this problem?



God designed an elegant solution to the problem. He designed the blood-brain barrier. The capillaries in the brain are different from the capillaries that are in the rest of the body. In our body, capillaries do not have any gaps between the cells, but the blood-brain barrier is a semi-permeable barrier that acts as a physical obstacle as well as a cellular transport system. It prevents harmful substances, such as bacteria, from entering the brain but allows the passage of essential nutrients. The cells that line these capillaries have special chemical processes to identify things that the brain needs (water, oxygen, glucose), and these processes transports them into the brain. Other molecules are recognized as bad for the brain, and they are kept out.

Isn't it amazing how our brain is protected in such a marvelous way? However, not all toxins are held back by the blood-brain barrier. For example, certain drugs can pass through the blood-brain barrier and wreak havoc on the brain cells. People who use drugs (including alcohol) looking for a "high" do not realize that they are actually circumventing the protection system that God has designed for our very sensitive brain cells. As a result, most drugs that affect the brain also kill brain cells. That is one reason the term "dope" has been used to describe illegal drugs.

God knew our brain cells were sensitive and needed protection; we can praise Him for creating a blood-brain barrier.

***Did not he that made me in the womb make him? and did not one fashion us in the womb?***

***~ Job 31:15***

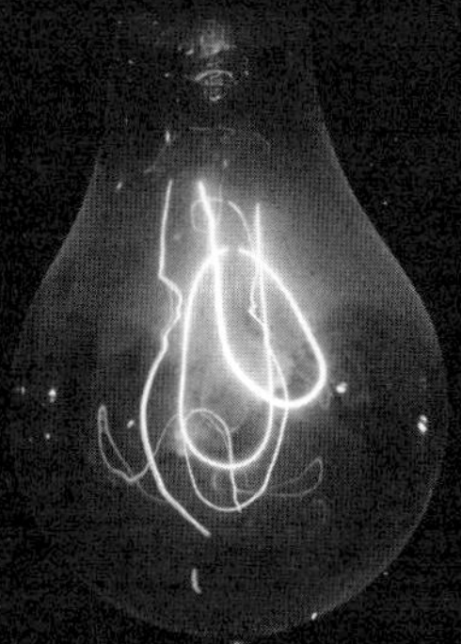

## INSPIRED BY PHYSICS — SEPTEMBER 6TH

Evolutionary theorist Dr. Herbert Spencer (1820-1903) spent many years promoting the idea that the universe made itself. After years of contemplating the make-up of the universe, he came to the conclusion that all physical reality could be described in terms involving just five basic components:

TIME
FORCE
ENERGY
SPACE
MATTER

Is it pure coincidence that the very first verse of the Bible contains these exact same five components? Moses penned the following words approximately 3,500 years before Dr. Spencer came to his conclusion:

In the beginning…{TIME}
God…{FORCE}
created…{ENERGY}
the heavens…{SPACE}
and the Earth…{MATTER}

We are given the privilege to study, understand, use, and have dominion over this marvelous creation. But God makes it absolutely clear, from the very first verse of His inspired Word, that He is the one who made this universe.

**In the beginning God created the heaven and the earth.**

**- Genesis 1:1**

Visitors to the southwestern United States are often awed by the imposing yucca plant. At its base is a rosette of stiff, sword-shaped leaves with a tall stem containing clusters of white, waxy flowers. The yucca plant can only be pollinated by one insect, the yucca moth, because the nectar glands can only be reached by the proboscis (sucking mouth part) of this moth. Likewise, the yucca moth requires the yucca plant for its reproducive cycle and for food.

When the moth visits the yucca flower, it collects pollen and carries the tiny pollen balls from plant to plant. After the female lays her four to five eggs in the yucca flower's ovary, she deposits her pollen ball on the tip of the flower's pistil, thus pollinating the yucca flower. The seeds then start developing at the same time the moth larvae develop. The seeds are the only source of food for the larvae. These seeds were made possible only by the pollen the female moth had earlier deposited. The larvae eat about half of the 200 seeds produced. The yucca plant could not survive without the yucca moth, and the yucca moth could not survive without the yucca plant.

If evolution is true, which came first? Both the yucca plant and the yucca moth had to be fully functioning from the beginning for this complex symbiotic relationship. God displays creativity in what He has made because He wants His existence apparent to all.

**I will teach you about the power of God.**

**~ Job 27:11a**

One of the most common criticisms of creation is the supposedly poor designs found in nature. In other words, if God made such and such a creature, feature, or organ, why is it designed so poorly?

This faulty logic assumes that all features of every animal still function exactly as they were designed to function at the beginning of creation. It is ignorance of a biblical worldview (often purposeful) which leads to many misconceptions being promoted as scientific facts. The incredible design and coded information of the DNA molecule would have been perfect at the time of creation,



with every part having some functional or designed purpose. However, at the time of man's rebellion, God allowed features and functions within creation to start deteriorating. Had God not done this, we would have lived forever, separated from Him. Thus, many features promoted as non-functional (such as blind eyes on cave creatures) are simply evidence for the decay of creation.

Furthermore, features promoted as poor designs inevitably turn out to be the result of a misunderstanding of the feature's function rather than the result of random mutational processes. Common examples include the belief that the Panda's thumb is poorly designed (it is a perfect design for eating bamboo – which is the Panda's favorite food) or the reverse design of our retina that has the blood supply blocking the light to the retina (this does not hinder vision but allows rapid reproduction of "blood-thirsty" retinal cells). This arrangement is also the optimum for producing a sharp image![1] It is pure human arrogance to believe that biological designs that we do not fully understand are faulty, as if we are capable of designing the creature to function more perfectly in its environment.

**God resisteth the proud, but giveth grace unto the humble.**

**- James 4:6**

Evidence for a Global Flood can be seen in the Grand Canyon.

- Layers are near-flat with knife-edge surfaces between layers. Unlike today's land surface, which is constantly being eroded by water and wind, rock layers are amazingly flat, showing little evidence of gullies and other erosional features.
- Layers over entire continents: Many of the rock layers at the Grand Canyon can be traced over vast areas of North America, Europe and the Middle East. Only a worldwide flood could carry and lay down sediments over such vast areas.
- Sand carried across the continent: Some of the sand and limestone in the Grand Canyon rock layers have been traced to the Appalachian Mountains. No river could strip away and carry such a sediment load and drop it out West—only a worldwide flood could do this!



- Layers of rocks nearly one mile deep are folded, not fractured: In portions of the Grand Canyon we find nicely folded rocks. How can this be? A flood would have laid down the layers in rapid succession, and then the layers were folded while still soft and pliable.
- Fossilized sea creatures far above sea level: Every continent in the world has fossilized sea creatures. Ocean waters once flooded over all the continents, carrying with them sea creatures.

The Grand Canyon's rocks are crying out that they were made by a worldwide flood of Biblical proportions.

**And behold, I Myself am bringing floodwaters on the earth, to destroy from under heaven all flesh in which is the breath of life; everything that is on the earth shall die. ~ Genesis 6:17**

Do you realize that some comets orbit backwards?

Comets are small "dirty balls of ice". Each time they pass by the sun, some of their ice evaporates. Comets lose so much mass every time they pass by the sun that they cannot be more than thousands of years old, otherwise they would have disappeared long ago. Recently, two comets have been found that have perplexed scientists. This is because their orbits are backwards around the sun, backwards in that the comets are orbiting in the opposite direction to the planets. It was so odd that when astronomers first discovered it, they ignored it thinking this observation had to be an error. When a second comet was found orbiting backwards, astronomers became convinced it was not an error.

## INSPIRED BY COSMOLOGY

SEPTEMBER 10TH

The reason this backwards orbit is baffling for astronomers is that they believe in the "nebular hypothesis". This evolutionary theory proposes that the sun, Earth and the rest of the solar system formed from a cloud or nebula of dust and gas. Such a collapsing cloud would have resulted in objects orbiting in the same direction. These two comets contradict this hypothesis by orbiting in the opposite direction. These backward orbiting comets are no accident; they declare their Creator. God wants to make it abundantly clear that He, not time and chance processes, is what created the universe.

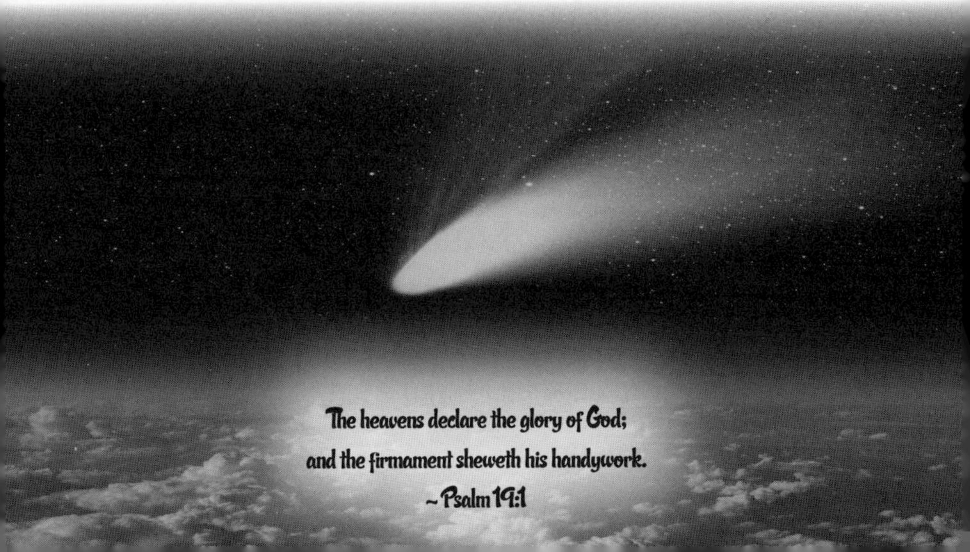

Nobody likes nasty, bloodsucking, disease-infected horseflies. Why would God create such an annoying creature? The male horsefly drinks nectar, but the female drinks the blood of horses and other mammals. The females need to obtain nutrients for their egg production. As the females move from animal to animal, drinking their blood, they often transmit disease causing microbes. God originally created everything "very good", so the original created horseflies would likely have dined on green plants. There is evidence to support this even today as some African horseflies feed on the sap of plants.





In the sun-baked African country of Malawi, the larva of the horsefly, genus Tabanus, displays a unique design. When the larva is ready to undergo metamorphosis (transforming itself from pupa to adult), it buries itself in wet mud. With time, the hot African sun dries the mud, creating cracks. These cracks can kill the pupating insect, so the horsefly larva prevents the crack from ripping through if by fashioning a protective cylindrical chamber while the mud is still moist. The larva tunnels down through the mud, moving spirally, round and round in an almost perfect circle. It makes a helical path until it reaches a depth of about 3.5 inches. Then the larva corkscrews its way back to the surface in an opposite spiral. At about ½ inch below the surface, the larva burrows into the cylindrical tunnel and seals it up. The larva is now encased in a reinforced mud cylinder. As a crack moves across the drying mud and hits the mud cylinder, the crack follows the curves around the cylinder – bypassing the pupating insect which is sealed inside. Normally the crack would cut straight across, but the larva compressed the mud in a circular pattern much like the rings of a tree. Did the horsefly larva learn this trick in order to survive? Hardly. God programmed the horsefly to encase itself in just this manner.

**Canst thou by searching find out God?**

**~ Job 11:7a**

You may have heard that the Bible is full of contradictions. Let's examine one such supposed contradiction.

In Leviticus 11:13-19, a bat is called a bird, "These are the birds you are to detest and not to eat because they are detestable...the eagle...and the bat." Everyone knows that a bat is not a bird! So is the Bible wrong? One needs to go back to the original language of the Bible, in this case, the Hebrew language. The Hebrew word for bird is owph which means "winged creature" or "fowl". Now replace the word "bird" with "winged creature". "These are the winged creatures you are to detest…eagle … bat." (Leviticus 11:13-19) A bat is a winged creature. There is no contradiction.

God's word is not full of contradictions, but a translation can be fallible. If you find an apparent contradiction in the Bible, go back to the original language and search it out. In this case, the translators did not pick the best word for owph. They are the ones that are fallible, not God. God's word is perfect and simply cannot contradict itself. No other book in all history can make such a claim and stand up to such scrutiny.

**All your words are true.....**

**~Psalm 119:160**

If the Tower of Babel existed, where is the tower? The actual structure is long gone, but many cultures from around the world



have historical legends that describe the Tower of Babel.

- The Polynesian island of Hao has a story about God coming down, tearing down a tall building, and giving the builders diverse languages.
- A Chaldean story tells of the inhabitants of the Earth who were proud of their strength and began to build a tower. God sent a wind that ruined their tower and introduced a diversity of languages; prior to this time they spoke the same language.
- The Kaska Indian people have a story called "History of Berosus" describing the location of Babel in Babylon on the banks of the Euphrates River.
- The Sumerian epic, "Enmerkar and the Lord of Aratta", also gives an account similar to the Tower of Babel.
- The Toltec Indians of ancient Mexico have recorded in their history that the first world lasted 1,716 years and was destroyed by a great flood. A few men escaped the flood in a closed chest. Following the great flood, the men began to multiply and built a very high tower. However, their language became confused, so different language groups wandered to other parts of the world. The Toltecs claim they started as a family of seven friends and their wives who spoke the same language. They crossed great water, lived in caves, and wandered for 104 years till they came to southern Mexico. The account reports that this took place 520 years after the great flood.

These are just a few accounts of the Tower of Babel. As people spread out, they would have brought their history with them. The Tower of Babel was a real event and is still the best explanation for the origin of the many languages of the Earth.

**And they said, "Let us build us a city and a tower, whose top may reach unto heaven; and let us make us a name, lest we be scattered abroad upon the face of the whole earth." - Genesis 11:4**

Fossilized thorns have been dug up in rock layers assumed to be 410 million years old by evolutionary paleontologists. Why do we have thorns so long before Adam and Eve existed? This is a huge problem for Christians who have compromised and believe in an extremely old Earth. According to the Bible, death, disease, bloodshed, and even thorns upon plants are the result of mankind's rebellion against God. These things exist because Adam and Eve sinned. According to evolution, man wasn't around until the last million or so years. Evolution places thorns before man, but God says man's actions resulted in thorns. These explanations cannot both be true…so which is correct? Throughout time, the Bible has been proven correct archeologically, theologically, and historically. These fossil thorns are not millions of years old; the age of rock layers they have been found in are simply being misinterpreted because the flood of Noah is ignored by evolutionists.

Fossilized thorns resulted from God cursing the world because of man's sin. Who can save us from this curse? Jesus Christ. In Matthew 27: 29, the Roman soldiers mocked Jesus by putting a scarlet robe on him and then "twisted together a crown of thorns and set it on his head." Here we see those cursed thorns again! This time the thorns are on the King of the Universe, God Himself. God took the curse of sin upon Himself and died in our place. Praise be to God for freeing us from sin! Thorns came after man sinned. Fossilized thorns are not millions of years old but are just evidence of a fallen creation that was trapped within the rock formed during the Flood of Noah.

***...cursed is the ground for thy sake...thorns also and thistles shall it bring forth to thee.***

***~ Genesis 3:17,18***

Loons eat fish and lots of them. Just like we need fishing gear to fish, the loon also needs gear. The loon's design for fishing includes swim goggles, swim flippers, and a special valve to prevent water from going down its throat.

- The loon's swimming goggles are a set of clear, built-in, third eyelids. These transparent eyelids (nictitating membranes) cover the eyes as the loon swims, giving it clear underwater vision.
- The loon's webbed feet act like swim flippers, so it can rapidly swim through the water to catch fish. The webbed feet are



positioned near the rear of the torpedo-shaped body allowing it to propel quickly through water. With the legs this far back on the body, walking on land can be awkward.
- Every time the loon tries to catch a fish for dinner, it opens its beak underwater. So why doesn't the loon get a mouth full of water? Loons have a built-in valve that closes when water is present – thus allowing them to catch a fish without swallowing a mouthful of water.

God thought of everything. God designed the loon with the right built-in fishing gear: goggles, flippers, and a throat valve.

The humpback whale has bumps on the front edges of its flippers. Common sense tells us that a smooth leading edge is most efficient. This is why swimmers shave their legs and wear tight, slick suits – to reduce drag in the water. This is why submarines are designed in a smooth, tube-like shape. Smooth skin was the conventional underwater design, until engineers actually tested a whale's flipper in a wind tunnel.

The smooth flipper behaved like a typical airplane wing – providing lift and propulsion to the whale in the water. Surprisingly, the flippers with bumps (tubercles) were found to have 8%



better lift and 32% less drag. It also resisted stalling (drastically losing lift) at a 40% steeper wing angle. Why did the bumps on the flipper work so much better? The bumps (tubercles) on the leading edge of the flipper broke up the flow of fluid and forced it into the fluted valleys in between. This generated eddies (vortices) that kept the flow attached to the top surface of the flipper, thus increasing the lift and resisting stalls.

Researchers are now proposing this tubercle design for use on helicopters, propellers and ship rudders. When we take time to study creatures, we are often surprised by all we learn from the design. We have much to learn from the Master Designer. It would be foolishness to attribute these intricate design marvels to chance processes such as random mutations guided by natural selection. When we see bumps (tubercles), we know there must be a bump-maker; and that is God.

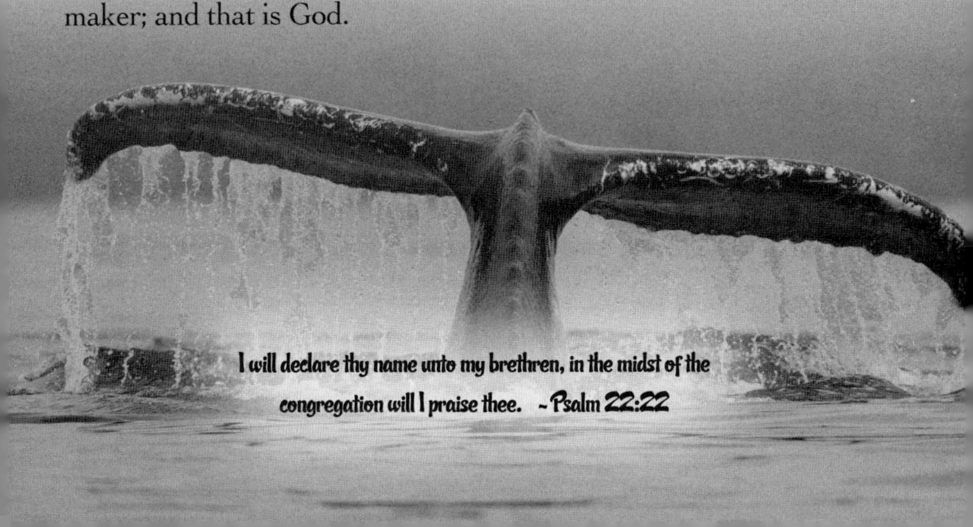

It was early 1900s, and Paul Sperry liked to go sailing. However, this hobby of sailing was tough, especially with the flat-soled shoes that he wore. Sperry would slip and slide on the wet deck, and that was dangerous. One day he noticed his cocker spaniel running on an icy winter path. Why wasn't he slipping and sliding?

Sperry examined the dog's paws. The pads had hundreds of tiny cracks and cuts going in all directions. Sperry found some rubber and cut grooves in a herringbone pattern. Then he attached this rubber to his flat bottom sailing shoes and went sailing. The improved traction was obvious on his wet deck.

Sperry went on to manufacture the first non-skid deck shoes. Today, grooved soles are a standard on sports shoes. Even many tire treads have a V-shaped herringbone pattern in order to provide



optimum traction on wet and icy roads. God has provided answers to many of the dilemmas we face. He is just waiting for us to search out His secrets.

**And let the beauty of the Lord our God be upon us: and establish thou the work of our hands upon us; yea, the work of our hands establish thou it.**

**~ Psalm 90:17**

Wind is air in motion, and this moving air has weight. This is why you can feel moving air. Wind moves from high pressure to low pressure regions. What happens when you stick a balloon with a pin? The high pressure air pushes outward through the tiny hole. High pressure moves into low pressure, and you can feel the wind (moving air) coming out of the balloon.

Wind blows because air molecules work to equalize pressure differences. So why and where do these different air pressure regions form? They are a result of the sun shining down and heating the Earth. Is the Earth heated evenly? No! The Earth's equatorial regions receive more solar radiation than polar regions. These differences in temperature result in pressure differences…which result in the creation of winds. Winds are created when the air tries to even out the temperature and air pressure differences.

Imagine Earth without wind; moisture would not be moved around, and many places on Earth would become parched. Temperatures in the Polar Regions would get colder, and those along the equator would get hotter, making the livable area of the planet much smaller. The Master Designer has made wind, so



moisture can be moved around, and Earth's temperatures can be moderated. If God did not design air with the perfect weight, then none of this would be possible.

*For he looketh to the ends of the earth, and seeth under the whole heaven; To make the weight for the winds...*

*~ Job 28:24-25*

The 2000 movie *Mission to Mars* (seen by over 20 million people and earning $110 million worldwide) had a plot centered on the discovery that aliens had sent DNA to Earth billions of years ago. Humans are shown to be the result of the evolution of this "seeded" biological life onto our planet. This idea, called pansperma, is currently acknowledged and debated amongst academic circles, in spite of the fact that there is absolutely no scientific support for this idea, and it is an unprovable belief rather than a scientific theory. Why is it even being considered among scientists as brilliant as Nobel nominee Sir Fred Hoyle and cosmologist Dr. Stephen Hawking? Because these brilliant scientists understand that life arising via natural processes here on Earth is an absolute impossibility.

There is no chance that even the simplest form of life could have developed from non-living chemicals. Every form of life requires specific proteins which are only produced from coded information found on the DNA molecule. DNA can only be made using proteins which only DNA can make. It is like trying to explain the existence of a zipper without the puller, which opens and closes the zipper. Suppose the information needed to construct the puller could only come from an unzipped zipper, but the zipper could never be unzipped without the puller. Obviously, both would have to be created simultaneously by an intelligent designer.

September 19th **MICROBIOLOGY** INSPIRED BY

The assumption of evolution blinds even brilliant scientists. Life could not have organized itself anywhere in the universe. Transferring the impossible to another planet does not make it more possible. The very existence of life can only be explained by a Creator of life.

**A scorner seeketh wisdom, and findeth it not: but knowledge is easy unto him that understandeth. Go from the presence of a foolish man, when thou perceivest not in him the lips of knowledge.**

**- Proverbs 14:6-7**

The adult male peacock has about 200 tail feathers, with 170 of those having the "eye" pattern. The eye feather has a precise pattern. The pupil of the eye feather is a dark purple, and the iris is blue. This "eye" is surrounded by bronze and green colors. How is this pattern made? These colors are not pigmented colors but are structural colors made by an optical effect called thin – film interference. You observe thin-film interference in the glistening blues and greens of oil slicks on a road. Oil is almost transparent; however, the light bounces off it producing the colors we see. Variations in oil thickness are the reason for different colors.

## INSPIRED BY BIOLOGY — September 20th

In the same way, peacock tail feather strands (barbs) are transparent, but the thickness varies. Light hits the transparent barbs and, depending on the thickness, reflects back a glistening color to create a beautiful pattern. A single barb can vary from blue, dark purple, bronze, and green because the thickness of the barb varies along the feather's length. Yet all the barbs coordinate with great precision to produce the eye pattern. It is amazing how each barb must be perfectly coordinated with every other barb to produce the "eye" pattern.[1] When we see such an intricate pattern and design made by such a complex mechanism, we know there must be a pattern maker, and that pattern maker is God.

It is also interesting to note how evolutionists try to explain such beauty. They say such beauty developed because male peacocks with the prettiest tails attracted the most mates. When this theory was recently tested, it was demonstrated to be totally wrong. Peacocks with plain or distorted tails attracted just as many (or more) mates than those peacocks with perfectly colored tails.[2]

**The Mighty One, God the Lord, has spoken and called the earth from the rising of the sun to its going down. Out of Zion, the perfection of beauty, God will shine forth. ~ Psalm 50:1,2**

Insect fossils have been found in the same layer as dinosaurs. These insect fossils look just like our modern insects. If evolution were true, we should find one creature changing into another in the fossil record. Yet, what we find in the fossil record are modern looking fossils of crickets, katydids, millipedes, spiders, and centipedes in the same rock layer as the dinosaurs. No change. A fossilized termite nest, on display at Petrified Forest National Park, looks just like a modern termite nest. The fossil record reveals that modern looking wood wasps, scorpion flies and mayflies flew with the flying reptiles (pterosaurs).

Did modern types of birds live at the same time as T-rex? Yes! Paleontologists have dug up bones of a parrot, penguins, and owls in the dinosaur rock layers. Also found have been fossilized bones of loons, albatross and ducks, along with shorebirds such as the avocet. Did flamingos and sandpipers live during the time of the dinosaurs? Yes, just like the ones living in Florida today! Have these insects and birds changed or evolved? No. This is called stasis (non-change) in the fossil record and is the opposite of what one would expect if

evolution were true. These creatures in the fossil record look essentially identical to their modern counterparts. God created these creatures during Creation week as we read in Genesis[1]. He created each creature "according to its kind." That is why we find fossilized creatures looking like modern creatures. Dinosaurs walked with the same creatures we see today.

**And Adam gave names to all cattle, and to the fowl of the air, and to every beast of the field;**

**- Genesis 2:20**

Why are almost all marine fossils found on the continents and not in rock layers on the ocean floor? Fossils of sea creatures are found at the top of all the major mountain ranges, even the 20,000+ foot Himalaya Mountains. Where should these marine fossils be? One would think they should be on the ocean floor; after all, evolutionists teach that the marine creatures have been living and dying in the ocean for millions of years. But, there is a stark absence of fossils on the ocean floor.

The biblical viewpoint of Earth history acknowledges a catastrophic flood with tsunamis that brought the marine creatures onto land surfaces. These ocean deposits flooded over the continents, carrying with them sediments and marine creatures. During the year long flood of Noah's day, there was not just one tsunami but many repeated disasters in such great numbers that



eventually thousands of feet of marine-bearing fossil sediments were laid down. When one visits the Dinosaur National Monument in Utah, **one finds fossilized dinosaurs mixed with marine fossils;** sea shells are the most common fossils in this "dinosaur deposit." The flood waters covered the continents, bringing with them the marine creatures. Throughout the fossil record (geological column), sea creatures are mixed with the fossilized land creatures. At the end of the year long flood, the mountains rose up and the ocean valleys sank down as the waters rushed off the continents into today's oceans. Apparently there was also rapid continental movement during this flood, which explains the location of the current continents and the formation of the new sea floor with relatively little sediment and few fossils. The flood of Noah's time was truly a violent cataclysm!

**And the waters decreased continually until the tenth month: in the tenth month, on the first day of the month, were the tops of the mountains seen. ~ Genesis 8:5**

The owl is a bird made to function at night. Every part of the bird seems designed for finding and killing prey.

- An owl has incredible night vision, which allows him to see even when there is no moonlight.
- With his incredible hearing, the owl is able to hear a mouse underneath two feet of snow! It has been said that, "If an owl can see you, it can hear your heart beating." During a walk in the woods, I once saw an owl watching me; it was a scary thought that this creature could hear my heart beating!



- Owls need to be quiet in their flight in order to not frighten off their prey. God designed an owl's flight feathers to be fringed, so it is a silent predator. Other birds have crisp edges, and this makes for a noisier flight.

If any of these parts (the night vision, super hearing, silent flight, talons and tearing beak) were missing, the owl would not be able to catch its prey. The forest would be overrun with rodents, and the owl would starve. If the owl had a robin's beak instead of the tearing beak, or crow's feet instead of talons, or a noisy flight like that of a goose – if just one feature was missing – the owl would not be able to find and kill its prey. Every part was needed for the owl to get its lunch (although these parts likely had other uses before the Fall). The owl was designed. When you see a design, you know there must be a Designer.

Have you considered what the Bible means when it uses the word "day"? In Genesis 1:5 it says, "And there was evening, and there was morning, the first day." What does day mean? In the Hebrew language, the word day is "yom", that can mean a 24-hour period or



a length of time (such as "in my father's day"). So which is it, a solar day or a long period of time? When we look at the phrase, we find descriptive words with day: evening and morning, and a number (2nd, 3rd, 4th, 5th, 6th). To the Hebrew reading this passage, it could have only one possible meaning: a literal 24-hour day.

Some Christians strongly influenced by an ancient Earth evolutionary perspective claim that the meaning of "yom" in Genesis cannot be known because Genesis 2:4 talks about creation in the "day of the Lord". They claim that this proves the meaning of "day" in the first two chapters of Genesis cannot be known. Consider the following sentence, "In the day of the civil war, Abraham Lincoln rose up early in the morning on day one of the memorial to give the Gettysburg address". Does anyone have any doubt which use of the word "day" means a period of time and which "day" means a literal 24 hour day? The same is true of Genesis 1 & 2. Those who want to add lots of time to Genesis discredit other parts of the Bible.

The definitive evidence comes from Exodus 20:8. This commandment states, "Remember the Sabbath day by keeping it holy … For in six days the LORD made the heavens and the Earth, … but he rested on the seventh day." God wrote these very words on a tablet of stone with his finger. God writes that He created in six days and rested on the seventh day, and we are to copy that pattern for our lives. When we use Scripture to interpret Scripture, it becomes clear that God created in six regular, 24-hour days and rested on the seventh day.

**These are the generations of the heavens and of the earth when they were created, in the day that the LORD God made the earth and the heavens.**

**- Genesis 2:4**

Have you considered the scab on your arm? Do you realize that it takes ten processes to form that scab? This is called the Clotting Cascade. In order for a blood clot to form, there must be ten specific individual chemicals reacting in a domino effect for safe healthy clotting. If one of these chemicals is missing or is in the wrong order, then a person would bleed to death. If the chemicals activate twenty minutes after the cut or if the clot does not cover the entire area, the result is again death. If the clotting continues without stopping, then the body's blood system would clot up, and the person would die.

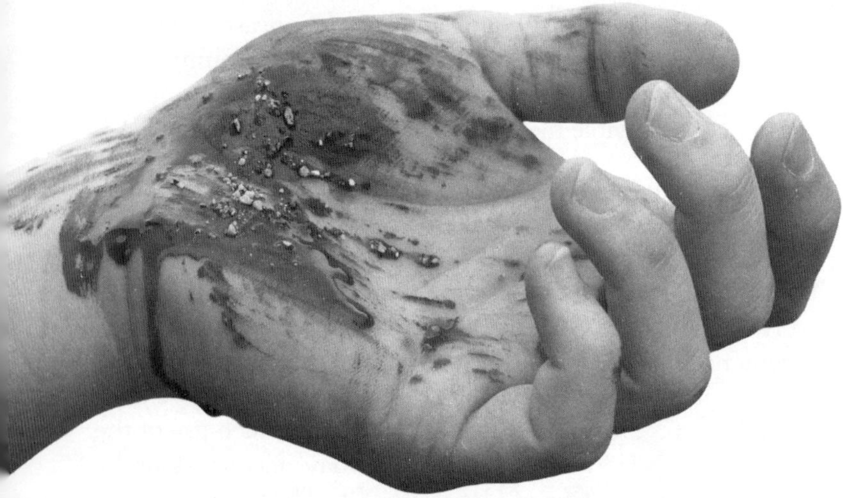

If evolution were true, how did this complex process happen by accident and chance? How did these chemicals get in the right order … in the right amount … at the right time? Our Master Designer



engineered this complex cascading process with all the chemicals, in the right amount, in the right order, and at the right time. Everything had to be present from the very beginning; otherwise, people would have become extinct long ago. This is a classic example of a principle of design called irreducible complexity. It is one of many scientific proofs that we must have a designer.

**Who hath delivered us from the power of darkness, and hath translated us into the kingdom of his dear Son. In whom we have redemption through his blood, even the forgiveness of sins. ~ Colossians 1:13,14**

A United States entomologist has estimated that one pair of Colorado potato beetles, if allowed to reproduce unchecked, would increase to over 60 million in one season. One female fly, beginning reproduction in May, would produce 143,875 bushels of flies by August. Aphids reproduce quickly, and in one season they can produce over 13 generations. If not held in check, the world's aphid population would be ten sextillion aphids in one year. Imagine living in a world where insects were not controlled. But what controls insect reproduction rates?

## INSPIRED BY BIOLOGY

SEPTEMBER 26TH

Fortunately, birds have a large appetite for insects. It has been found that a scarlet tanager can eat over 600 gypsy moth caterpillars in 18 minutes. It has been estimated that chickadees alone eat over 8 billion insects per year. Birds keep the insect population in balance. Now consider what evolutionists say about the origin of birds – they did not evolve until many millions of years after insects appeared on our planet!

Without birds, insects would have decimated the world's vegetation. The world would have been a bleak place with little life. What we see is actually what God has said, that He created both birds and insects during the first week of creation at the very beginning of time.

**And God blessed them, saying, Be fruitful, and multiply, and fill the waters in the seas, and let fowl multiply in the earth. - Genesis 1:22**

Did you know that starfish, sea urchins and sea lilies lived at the same time as the dinosaurs? So what, you ask? These creatures, echinoderms, are very different from each other, yet have little difference from similar creatures today. These fossils have been found in the same rock layers as dinosaurs, yet they have been given different names from *"modern"* echinoderms. This seems to be an effort to disguise the fact that evolution hasn't happened. When we compare, side by side, the fossil with its modern version, we marvel at their similarity. Lobster fossils, found with Archaeopteryx, look just like the modern Maine lobster with its large claws. Shrimp fossils from the dinosaur times look just like the shrimp we love to eat!

When a fossil is compared side by side with the living, modern day creature, we find it to be uncannily similar! No evolution happened. Even though fossil creatures are frequently given different Latin names than their modern counterpart, the creatures look essentially identical. Names given to creatures are very subjective, and frequently there are disagreements among scientists regarding how to classify animals and plants. Many of the same creatures

which lived with the dinosaurs are still living all around us and have been around since the beginning of creation.

Although many creatures have become extinct, we see no completely new creatures or complex features forming. Have creatures evolved? If so, we should find it in the fossil record. We simply do not.

**The thing that hath been, it is that which shall be; and that which is done is that which shall be done: and there is no new thing under the sun.**

**- Ecclesiastes 1:9**

The original documents of the Bible are no longer with us, so how do we know that the translations that have been handed down are correct? Manuscript studies is the field of knowledge that deals with these types of questions.

First, experts date the copies to determine how long it has been from the original writing to the copies in hand. Second, researchers evaluate the total number of copies available. The greater the number of ancient copies we have, the more accurate the record. Let's compare the New Testament with some other works of antiquity.

- Caesar wrote a history of the Gallic wars around 50 B.C. There are only 10 good copies, the oldest being 900 years after Caesar died.
- There are seven manuscripts of Plato's Tetralogies, with the earliest manuscript dating 1,250 years after the original.
- For the New Testament, there are over 24,633 manuscript copies, all within 100-150 years of the originals.

The Bible is by far the best-documented book in the history of mankind with more manuscripts and earlier manuscripts than any other classical work. There is not even a close second. The scholar



John Warwick Montgomery concludes, "To be skeptical of the resultant text of the New Testament books is to allow all of classical antiquity to slip into obscurity, for no documents of the ancient period are as well attested as the New Testament."

There is absolutely no reason to believe we do not have an accurate and reliable copy of both Old and New Testament scriptures.

***Your word is a lamp unto my feet and a light unto my path.***
***- Psalm 119:105***

Back in 1909, a Canadian named Jack Miner set out to discover to where birds migrate. He was the first person to begin banding ducks and geese in order to track their movements. Some years earlier, he had created a bird sanctuary on the north shores of Lake Erie where birds were numerous and could be easily captured. He decided to put his name and address onto ankle bands with the hope that distant hunters and other naturalists would write and tell him where the birds were shot. Sure enough, he soon started receiving letters revealing that the birds were migrating far to the south in the winter and far to the north in the summer.

In 1914, he began stamping Bible verses on the bands, making the birds "winged missionaries." Miner said, "From the very first time I stamped such a verse on a band, I felt the help of God and knew I now had my tagging system complete." Over the years, hunters wrote to Miner saying that the Bible verses had changed their lives. Eventually, a missionary from the remote region of Northern Canada came to visit Miner. The missionary stated that a revival had taken place among the Eskimos because of the Bible verses placed on the bird bands!

The next time you see a flock of geese heading south consider Jack Miner, a man who used his love of birds to glorify God. With God's help, how can you use your passions to lead others to Jesus Christ?

*Delight yourself also in the Lord,*
*and He shall give you the desires of your heart.*
*~Psalm 37:4*

Neptune is an enormous gas planet, the eighth planet in our solar system. This planet appears as a small bluish dot in our most powerful telescopes. Our best photos of Neptune were taken by the Voyager II spacecraft when it flew by the planet in August 1989. Based on evolutionary assumptions of an extremely old solar system, astronomers thought Neptune would be a cold, inactive place. But it was not.

Neptune has the strongest measured winds in the solar system, estimated at 1,300 mph. Its raging winds cause Neptune to have a dynamic atmosphere with two large spots, thought to be enormous atmospheric storms. The larger storm is about the same diameter as the Earth! To the amazement of evolutionary scientists, the Hubble Space Telescope revealed that these storms had dissipated in 1994 – again revealing an extremely active environment upon this planet.

Evolutionary theory predicted that Neptune should be cold and dead. Instead, it was found to generate heat, radiating twice as much heat into space as it receives from the Sun. This fits well with the Creation model; a young Neptune would still be cooling off.

Evolutionary scientists believe the solar system formed from an enormous cloud of gas and dust, with the gas giants forming at the outer edges. With this model, the more distant a planet, the longer it would take to form. According to evolutionists, Neptune should not even exist; it is too far out to have formed. Yet Neptune does exist, with its raging winds, dynamic atmosphere, and heat generation – all of which contradict an ancient evolutionary origin.

Neptune simply appears to be quite young. Perhaps it is because some 6,000 years ago, Neptune was created by God on Day 4 of the creation week.

**When I consider thy heavens, the work of thy fingers, the moon and the stars, which thou hast ordained; what is man, that thou art mindful of him? and the son of man, that thou visitest him? ~ Psalm 8:3,4**

# OCTOBER

*"Those who desire to give up freedom in order to gain security will not have, nor do they deserve, either one."*

**- Benjamin Franklin** (1706-1790)
Author, diplomat, inventor, physicist, politician, & printer.

*And ye shall know the truth, and the truth shall make you free.*

**- Matthew 6:33**

Have you considered the bud of a beech tree? This tiny bud, less than one inch long, opens to produce a leaf which is six inches across. How can such a big leaf come from such a tiny bud? It is folded in a precise way to allow maximum packing efficiency without the folds interfering with each other during opening.

This is what Japanese scientist Koryo Miura copied when challenged to send a solar panel into space. Miura copied the way a beech leaf was folded in the bud. He folded a massive 80 foot long solar array with a surface area of 620 square feet. This array was then stored in a rocket and sent into space. Upon its arrival, the compact solar array unfolded easily and without damage. This type of folding was so successful that it is now called the Miura-fold. The

Miura-fold is being used everywhere from Tokyo subway maps to heart stents that clear blocked arteries.

The Miura-fold should really be called the God-fold—for He was the one that first designed it. Miura just mimicked the design. God wants us to see His design in nature. Each spring, go



and look at a bud, and observe how God gets such a big leaf into such a tiny bud. See how it unfolds. God takes joy in our use of His designs to solve problems, especially when we acknowledge Him as the Great Designer.

**Let the field be joyful, and all that is therein: then shall all the trees of the wood rejoice.**

**~ Psalm 96:12**

Why are red blood cells disk-shaped with a bi-concave center? Why aren't they shaped like a sphere or a regular flat disk? A group of scientists decided to see if this bi-concave disk was the best shape for red blood cells, so they fed a number of parameters into a computer to see what results would be produced. The primary purpose of a red blood cell is to carry oxygen, so they used the laws of gas diffusion to see what shape was most efficient for this oxygen absorption. They came up with the following equation:

$$\int^{\pi/2} \frac{dx}{\sqrt{1-k^2\cdot\sin^2 x}} = \frac{\pi}{2}\left[1+\left(\frac{1}{2}\right)^2 k^4+\left(\frac{1\cdot3}{2\cdot4}\right)^2 k^4+\left(\frac{1\cdot3\cdot5}{2\cdot4\cdot6}\right)^2 k^6 \cdots\right]$$

The computer then drew the shape of the red blood cell using these fundamental laws of science. To their surprise, the computer drew a bi-concave disk, the exact shape of a red blood cell! But, of course, God designed the laws of science, so the best shape for a red blood cell is the way God designed it.

The disk's smooth, round edges reduce the amount of friction on the cell as it moves through thousands of miles of blood vessels. The thin center and thicker rim (bi-concave) allow the red blood cell greater flexibility, so it can fold as it moves through very narrow capillaries. If the red blood cells were spherical, they could rapidly absorb oxygen near the surface, but the process would slow significantly deep within the interior of the spherical shape and not be able to quickly satisfy the body's demand for oxygen.

## INSPIRED BY MICROBIOLOGY

OCTOBER 2ND

Our red blood cells have the optimum shape needed for flowing through our blood vessels and for chemical reactions to occur with speed and volume. When we examine even the tiniest parts of our body, we see design.

**...for the blood is the life...**

**~ Deuteronomy 12:23**

By the 1960's, Christianity was in full retreat in the United States. Bible colleges and seminaries were turning out pastors



poorly prepared to defend the straightforward understanding of Genesis. Thus, everything the Bible says about biology (creatures reproduce after their own kind), geology (a worldwide flood would have rapidly created most of the sedimentary rock layers of the Earth), and human history (death exists as a direct result of human rebellion) was not believed, mentioned, or defended from most pulpits. Not surprisingly, since the Bible was no longer considered relevant on issues dealing with the physical world, there was little outcry when it was removed from public schools in 1961. Outlawing of prayer rapidly followed in 1962.

The turning point in this situation came from an unlikely source. A little-known hydraulic engineer named Henry Morris (1918 - 2006) teamed with theologian John Whitcomb to produce a book describing the geological effects of a worldwide flood. *The Genesis Record* is credited with slowing the disregard for God's Word, and is considered the starting point for the modern creation movement and a tool used to bring tens of thousands of people to a real living faith in Christ. It was the denial of the Genesis flood (a century earlier) that had added millions of years to modern man's understanding of history and, thereby, destroyed the belief in the Bible as a trustworthy book. So, it is not surprising that a book confirming the reality of this flood was the turning point for restoring trust in God's Word.

Dr. Morris went on to found The Institute for Creation Research (ICR) - the premier Biblical creation research scientific organization in the world. He also wrote more than 20 books, which were distributed in dozens of countries over 50 years. Throughout these activities, he was insulted, disparaged, and ridiculed by both atheist organizations and mainstream denominations. Yet, his legacy lives on in the lives of transformed Christians equipped with the truth.

**...The effectual fervent prayer of a righteous man availeth much.**

**- James 5:16**

# INSPIRED BY BOTANY

OCTOBER 4TH

Have you considered how trees close up shop for the winter? Every autumn, all over the colder parts of the world, trees get ready for winter. All through spring and summer, their leaves have acted like little factories, busily making food. They absorbed carbon dioxide from the air, drew water from the soil, and used sunlight in a complex process called photosynthesis – allowing them to store food in the form of starches and sugars. But the onset of winter means freezing temperatures. Frozen leaves do not work well. Frozen ground water does not flow.

In the winter, food production stops, and trees close down. During autumn, the leaves on the trees turn from green to varying shades of yellow, red and orange. At the point where each leaf stem is attached to a twig, special layers of cells grow – making a wall. This blocks the water flow to the leaf, revealing colors hidden during the spring and summer by the green chlorophyll of photosynthesis. This wall also seals the gaping hole in the twig when the dead leaf falls off. Because the leaves have stopped needing water, the roots stop pulling water from the soil. By the time winter arrives, the tree's water transfer "pipes" have been drained, preventing internal damage to the tree from water expansion during freezing.

The tree also needs to be ready for the next spring, so even before it shuts down for the winter, while it is still actively growing during the summer, it starts forming next year's leaves. These leaves are packaged as neat little buds with weather-resistant scales in order to survive the harsh winter. When spring arrives, the tree is ready to reveal its hidden miniature leaves. How could anyone believe that this complicated process happened by accidental, random, chance, or mutational changes over time?

**For thou, Lord, hast made me glad through thy work: I will triumph in the works of thy hands. O Lord, how great are thy works! ~ Psalm 92:4-5**

When visiting museums we often see dioramas of cavemen, humans shown as brutish, hairy ape-men. In fact, people have



lived in caves throughout history; some even live in caves today. If you visit the town of Coober Pedy in South Australia, you will find people living in caves with TV antennas sticking out. This is an opal mining town, and the ambient outside air temperature is so high that some miners choose to live in caves where the temperature is cooler. Are these miners primitive, brutish ape-people? Hardly, they were just hot! People assume, based on evolutionary ideas, that humans who once lived in caves were primitive brutes.

After the Tower of Babel when people scattered throughout the world, some sought shelter in caves as they built more permanent homes. In Genesis, Lot lived in a cave after he fled Sodom. When King Saul was after David, David lived in a cave. Obadiah hid a hundred prophets in a cave in order to save them from Queen Jezebel. Elijah himself lived in a cave.

On the island of Malta, in the Mediterranean Sea, the tradition of living in caves goes back to antiquity. It was written about during the Middle Ages. The writer found the children strong and simply dressed. The adults greeted him. The women were remarkably good looking. The men worked the fields, tended animals, and took the produce to town for sale. The women made cheese and bread. Drinking water was stored in earthen pitchers. The caves were ventilated with shafts and decorated with crosses and holy pictures. Brutish, hairy ape-man? Throughout history people have chosen to live in caves (including large cave villages still in existence in China), but this does not mean they were primitive.

**So God created man in his own image, in the image of God created he him; male and female created he them.**

**~ Genesis 1:27**

Do you have bird breath? Not a chance! Birds do not breathe in and out like we do but have a different breathing system. Bird lungs are rigid and act like filters. Air flows through the lungs to air sacs that are sandwiched between muscles and limbs and act as bellows to move the air. It is complicated to envision, but basically, birds breathe in and out at the same time! Oxygen is rapidly absorbed as it flows against huge numbers of tiny blood vessels in the lungs. Maximum oxygen is obtained, and maximum carbon dioxide is quickly removed. It is because of this design that birds can fly vigorously for great distances without tiring.

## INSPIRED BY BIOLOGY — OCTOBER 6TH

The Blackpoll Warblers fly non-stop for the entire 85 hours it takes to migrate all the way to South America. A bar-tailed godwit, fitted with a tracking device, flew 7,100 miles non-stop in just over 8 days! Hummingbirds migrating from Minnesota meet their greatest challenge when they cross the Gulf of Mexico. They must fly 600 miles non-stop across the Gulf waters to reach South America. They do this in 18 hours. We think it's an accomplishment to run a 26 mile marathon in a few hours! God has designed an amazing way for birds to breathe, so they can be wondrously efficient flying machines.

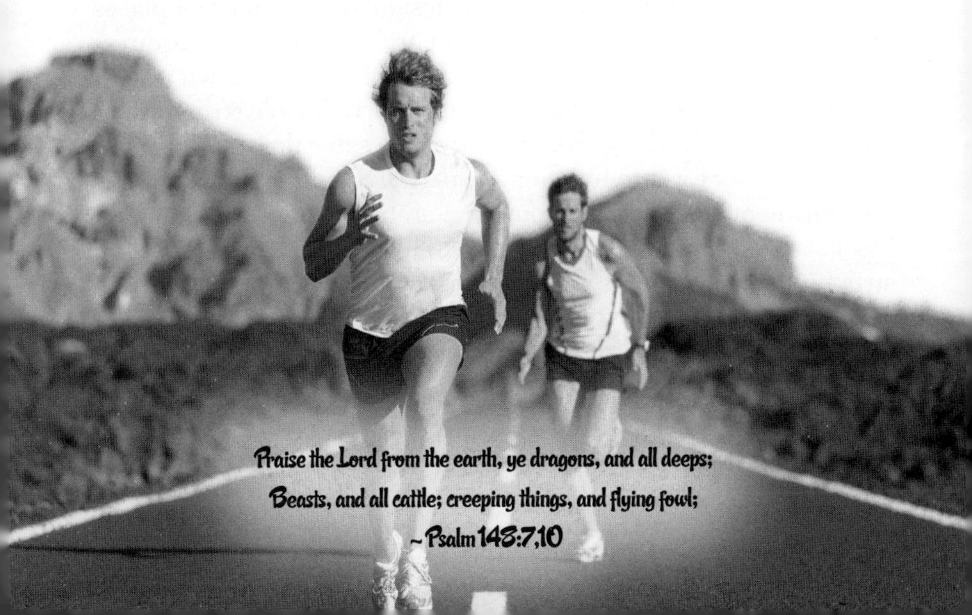

Have you considered the human foot? It has a unique arch that allows man to stand upright. What does it take to make an arch? Think of the Gothic churches of the Middle Ages and their towering arches. Arches need wedge-shaped blocks, a keystone and ties. The human foot has 26 precisely shaped bones, several which are wedge-shaped. The ankle bone is the keystone. As the weight of the body is pushed down on the keystone (ankle bone),



the weight passes through the wedge-shaped bones to the floor. The wedge-shaped bones are held in place with ligaments (ties). The human foot needs all three parts in order for an arch to be made. If one part is missing, the arch collapses.

In the human foot, the ligaments hold the bones together with enough stiffness to make the arch hold the weight of the body, yet flexible enough to allow the foot to flex for walking and running. This flexibility helps to absorb the shocks during walking and running. In addition, the muscles and tendons in the precise places enhance the arch allowing for wonderful balance in an upright position. A human's arched foot is the perfect design for walking and running upright. In contrast, ape feet have no arches – they use their feet as a second pair of hands for gripping branches.

Arches in buildings need differently designed parts in exact places before they can function properly. An arch in a building requires design and planning to work successfully. It is well known in engineering that an arched structure is an irreducible structure. All the pieces must be designed, present at the same time and placed in a correct order for it to function properly. Since the human foot is an equivalent (yet far more complex) arch utilizing wedge-shaped blocks, a keystone and ties, the human foot must be an irreducible structure too. All pieces must be present at the beginning for the arch to work. When we see an arch, we know there must be an arch maker. With the foot, this arch maker is Jesus.

**Lo, this only have I found, that God hath made man upright; but they have sought out many inventions.**

**~Ecclesiastes 7:29**

The Great Unconformity is a distinct, essentially flat surface, often separating the lowest layer of sedimentary rock (laid down under water) from upper layers of metamorphic rock (compressed, melted fused layers). This is called the Great Unconformity because this distinct line between rock layers extends across the entire North American continent and represents five hundred thousand to over a billion years of "missing time."

How could over a billion years of sediment have been removed over an area the size of the United States, leaving essentially a flat, eroded surface over the entire area? Where are similar widespread erosional processes happening ANYWHERE on our planet today? None exists. Yet during a world-covering flood such a process is conceivable as enormously energetic waves of sediment-laden fluid scoured the lower laying surfaces over enormous regional



areas before dropping loads of sand and other sorted sediment onto the newly sheared-off surface. This Biblical model for geology elegantly explains the great unconformity but also removes the possibility that there is any "missing time." Evolution needs this "missing time" for life to evolve. Textbook writers and geologists simply ignore the obvious implications of the Great Unconformity because they are trained to interpret everything via natural causes and need huge time periods to justify their conclusions.

The Great Unconformity fits perfectly into the Biblical model when the worldwide flood of Noah's day is factored into Earth history. It is really not an *unconformity* at all because it perfectly conforms with the correct view of Earth history.

**The stone which the builders refused is become the head stone of the corner.**
**This is the Lord's doing; it is marvellous in our eyes.**
**~Psalm 118:22, 23**

Have you considered the rivers in the sea? One of them, the Gulf Stream, carries 5,000 times as much water as the Mississippi River. We know that the warmth of this river keeps Great Britain from having unbearably frigid winters. There are many other "rivers in the sea." If they did not exist, the oceans would be stagnant and become filled with death and decay.

According to Jonathan Weiner in the book *"Planet Earth",* without this taking place,

> "noxious fumes would slowly gather near the floor and kill all the bottom dwelling creatures. Slowly, the Earth's own body heat would warm this foul black water until it was hotter than the water on the surface. The unstable ocean would eventually turn over and vent the whole mess to the surface, with catastrophic effects on fauna and flora, poisoning the atmosphere". [1]

In addition, charting these paths of moving water made ocean travel by early sailing vessels more efficient and the movement of people and goods across the globe a reality – long before modern mechanized shipping was developed. A great Christian pioneer of ocean navigation, Matthew Maury, first discovered these pathways in the sea based on a clue found in Scripture. Maury is still

## October 9th BIBLICAL ACCURACY INSPIRED BY

honored as one of the founders of the science of oceanography – all because he trusted God's Word to mean what it said.

Our life on Earth depends in many ways upon these rivers in the sea. God set them in motion for our benefit and then made sure we would find them by mentioning them in His inspired revelation to us – the Holy Bible.

**Thou madest him to have dominion over the works of thy hands... and the fish of the sea, and whatsoever passeth through the paths of the seas..**

**- Psalm 8:6,8**

Imagine a lake filled with fish, plants, and other creatures. In the winter, ice forms on the top of the lake. What if the ice sank to the bottom? And then another layer of ice formed on top and this too sank to the bottom? Eventually the lake would fill with ice, killing all the creatures living in it. Imagine this lake filled with ice, from the bottom up; how long would it take for the summer sun to melt it away? Depending on the locations of the lakes, many would only partially thaw during the summer. Imagine an ocean with ice caps, once they had formed, they would also sink. Eventually in that area, the ocean would fill up with ice.

Thankfully, when water freezes it does not sink but floats. This protects the animals living in the water below. Most liquids become heavier and sink when frozen but not water. Frozen water floats. As water cools to 39°F, it contracts; then a strange thing happens from 39 to 32°F. It expands until it freezes. This expansion creates spaces between the molecules causing the ice to float. Ice floats in water because it is less dense than water. Water expands 9% when it freezes, unlike almost all other liquids. Because of this anomaly, ice that forms on lakes, seas, and oceans



stays near the surface rather than sinking and causing destruction of life. Our Creator gave water this unique quality in order that life would be possible here on Earth.

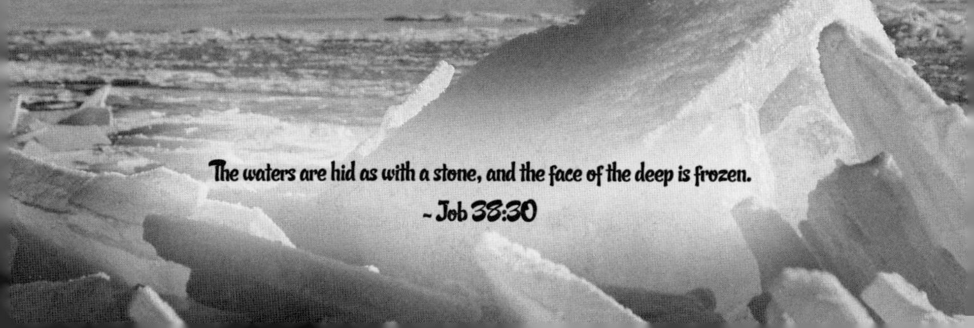

**The waters are hid as with a stone, and the face of the deep is frozen.**

**~ Job 38:30**

Have you considered how well archaeological evidence provides support for the recent creation of mankind? All anthropologists



know that key human technological advancements – agriculture, the wheel, pottery, and horse domestication – were developed less than 10,000 years ago. The oldest plow is about 4,000 years old. It was discovered in Lavagnone, Italy in a peat bog. What about agriculture in general? Archaeological excavations have shown that agriculture has existed less than 10,000 years. What about the wheel? Even secular history books admit the wheel was invented 5,000 - 6,000 years ago and not in just one area but in Germany, Switzerland, Slovenia, and Mesopotamia. What about pottery? The earliest pottery found was in Iran, Japan, and China and has been dated less than 7,000 years old. Some of the pottery was found to be just as sophisticated in design as recent pottery. What about horse domestication? Again, the most ancient date advanced is that horses were domesticated less than 6,000 years ago. Archaeological evidence shows that the three most ancient civilizations, the Egyptian, Mesopotamian, and Indus civilizations, arose 6,000 (or less) years ago.

Even the oldest written languages date back about 5,000 years, and these languages show no evidence of evolving in complexity. In fact, the most ancient languages were just as complex and sophisticated (often more so) than modern languages – having complex grammar and structure. When we examine the evidence of human activity through archaeology, we find overwhelming support for a recent creation of man. If man had been evolving over millions of years, we should find evidence to support this, but we do not. What we find is man leaving his evidence dating back far less than 10,000 years. This is exactly what we would expect by reading the Bible as accurate, literal history.

**For enquire, I pray thee, of the former age,**
**and prepare thyself to the search of their fathers:**
**~ Job 8:8**

## INSPIRED BY COSMOLOGY

October 12th

Earth's atmosphere is made up of a mixture of gases that perfectly support plant and animal life. Perhaps just as importantly, our atmosphere acts like an invisible blanket protecting us from the extremes of hot and cold temperatures. Contrast this with Mercury where there is no atmosphere; nights are freezing and days are burning hot. Venus' atmosphere is too thick, keeping the heat in, causing extremely hot days and nights. Earth's atmosphere moderates our temperatures.

A third benefit of the Earth's atmosphere is often forgotten - it protects us from rocks falling from the sky. Each day the Earth has rocks of every size flying into it. Imagine going outside and having large rocks, little pebbles and tiny grains of sand constantly pelting you. You would need a steel umbrella! Plants would be damaged and our buildings would crumble. But no need to fear… our perfect atmosphere burns up almost all of this debris. Very rarely does a rock make it through to the ground. When it does, the "rock" is called a meteorite.

The Earth has the perfect atmosphere for life.

**…that the Lord cast down great stones from heaven…they were more which died with hailstones than they whom the children of Israel slew with the sword.**

**- Joshua 10:11**

Spiders are capable of producing one of the finest filament strands known to man. These threads can be 10,000 times thinner than a strand of human hair, yet the silk is five times stronger than an equivalent weight of steel cable. Scientists have yet to learn to synthesize an equally strong artificial silk, nor do they know how a spider keeps from clogging its spinnerets as the emerging silk immediately solidifies upon exposure to oxygen. How could some "pre-spider" have just happened upon the correct formula for producing silk and, at the exact same time, evolved both the equipment for dispensing it and the instincts for how to use it?

## October 13th

## Biology Inspired By

We have also made some incredible discoveries from spider silk. Scientists recently coated silk with a type of glass $SiO(C_2H_5)_4$ and then burnt away the inner silk in a 788°F oven. The resulting glass shrinks fivefold leaving a hollow glass tube with an interior diameter 50,000 times smaller than a human hair. These tubes can be used as light pipes for light-speed optical circuits or as nano-sized test tubes for single molecules of specific chemicals.

In another discovery, an Israeli company has produced a large spider web shaped net of fine fibers in the shape of an inverted tent. At night, dew droplets form on the fibers in the same way they collect on a spider's web in the morning. The droplets roll downwards into a collection container; a 20 foot diameter web can collect up to 12 gallons of water per night from desert air.

What marvelous wonders we can learn from God's creation.

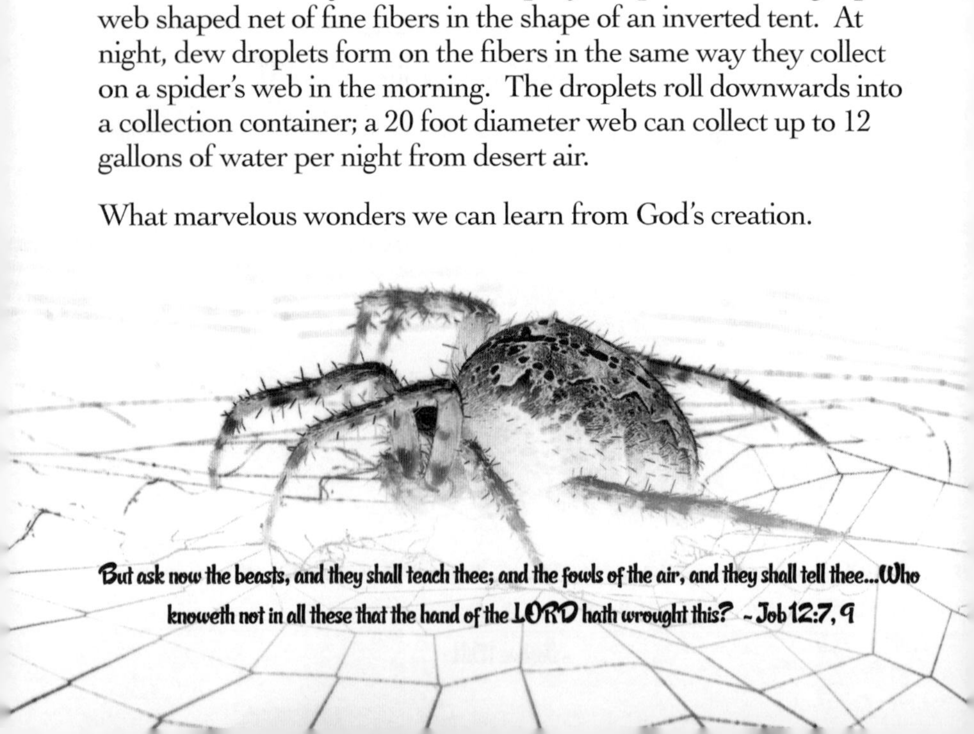

***But ask now the beasts, and they shall teach thee; and the fowls of the air, and they shall tell thee...Who knoweth not in all these that the hand of the LORD hath wrought this? - Job 12:7, 9***

There have been hundreds of thousands of dinosaur tracks found all over the world - from Africa to Australia, Europe to the Americas. The Puragoire Valley in southeastern Colorado alone contains 1,300 tracks. Just outside Denver, Colorado is Dinosaur Ridge with some 350 dinosaur tracks. St. George, Utah has dinosaur trackways, tail drag impressions and one place where a dinosaur actually sat down in the mud! The Paluxy River near Glen Rose, Texas contains numerous large dinosaur footprints, some so large a young child can sit down inside the track. The tracks are also only found on flat "bedding planes." So how do we fit these observations into a biblical worldview of history? It begins with Noah's flood.

Many dinosaurs apparently fled the rising flood waters which had already washed over low lying areas of the ever-changing terrain upon the Earth. New layers of sediment, at times thousands of feet thick, had been washed into various places upon the planet early in the flood while mobile animals, such as dinosaurs, attempted to

## INSPIRED BY PALEONTOLOGY

survive the ever rising flood waters. During their attempted escape, many left impressions upon the newly formed flat bedding planes. These tracks were subsequently covered with hundreds of feet of sediment laid down on top of the tracks. These sediments were later eroded away, leaving the impressions of dinosaur tracks for us to find.

Today our footprints from walking on a beach or through a muddy field are quickly obliterated. To preserve dinosaur tracks would require rapid, deep burial. To have footprints preserved worldwide would require a worldwide flood such as is clearly described in Genesis 6-9.

**Thou rulest the raging of the sea: when**
**the waves thereof arise, thou stillest them.**
**- Psalm 89:9**

## October 15th



Look closely at the shape of deciduous trees (those that lose their leaves in the winter) silhouetted against the winter sky. Each type of tree has its own characteristic canopy shape. This is as a result of the way the branches grow. Why do the branches grow this way? Each leaf is like a miniature solar collector, bringing in the sun and combining it with water and nutrients to make the tree's food. Each leaf has to have the maximum possible sunshine touching it with the least amount of shade. The tree branches hold these mini-solar collectors (leaves) in an optimum position depending on the leaf's size and shape. Each type of tree has different leaf shapes and sizes that require the arrangements on the branches to be different. This requires a different canopy shape for each tree type. Trees need sunlight to grow; without it they wither and die. So God designed a tree's shape in order for the tree to receive the resources it needs.

**And he shall be like a tree planted by the rivers of water, that bringeth forth his fruit in his season; his leaf also shall not wither.**

**~Psalm 1:3**

Scientists agree that James Clerk Maxwell's (1831-1879) greatest contribution to science was his brilliant 913 page treatise on the mathematical unification of electricity with magnetism. This forever changed the world of physics and is the foundation of radio transmissions, television, computers, space travel, and all modern electronics. Maxwell stands at the forefront of the most brilliant scientists who ever lived, and his legacy is the modern world we all enjoy. Some physicists have shown that even atoms can be explained as the result of the movement of charges and that matter itself is really an electrical simulation – all based on Maxwell's equations.[1]

Although Maxwell lived a short 48 years, he was also responsible for mathematically explaining Saturn's rings as a massive cloud of particles. He made major contributions to the fields of statistics, astronomy, gas theory, color theory (producing the first color photograph), and thermodynamics.

As a contemporary to Charles Darwin, Maxwell soundly rejected naturalistic explanations for life stating, "No theory of evolution can be formed to account for the singularity of molecules, for evolution necessarily implies continuous change…"

A prayer written by Maxwell stated, "Almighty God, who has created man in Thine own image, and made him a living soul that he might



seek after Thee and have dominion over Thy creatures, teach us to study the works of Thy hands…"

A deeply sincere follower of Jesus Christ, Maxwell used his faith as the foundation from which he made his great discoveries. Do not be deceived that belief in creation is a detriment to scientific advancement … the very opposite is true.

Maxwell had this verse inscribed upon the entrance to his science laboratory at Cambridge University:

**The works of the Lord are great,**
**sought out of all them that have pleasure therein.**
**~ Psalm 111:2**

Bats rely on echolocation for hunting in the dark. The bat opens its mouth and sends out a high-pitched sound, which humans cannot detect. When the sound hits something, it bounces back making an echo, which the bat can hear.



To make high-pitched sounds, a bat has a specialized larynx that makes intense, high-frequency ultrasounds. From these ultrasonic sound pulses and their echoes, bats can determine the distance, size, shape, surface texture and speed of their prey.

A bat's ears collect these ultrasounds. Here enters a problem – the sound pulses emitted are powerful, and the bat's ears are sensitive. So how are the sensitive receptors not overwhelmed by the outgoing sound waves? Humans had to deal with this problem when they developed modern radar. The transmitter would send out powerful beams of radar that were too powerful for the sensitive receiver. The problem was solved by developing a fast switch that turned the sensitive receivers off every time a radar pulse was sent out. That is exactly what was found in bats!

When the bat is sending out its high-pitched signals, muscles in the ears close for a split second, thus protecting the sensitive ears. With the two specialized features of a transmitter (larynx) and receiver (ears), the bat then must have the ability to analyze what these sounds mean. Within a few thousandths of a second, the bat has built a mental image of its surroundings - much like we do with our eyes. The bats ability to "see" using sound waves is irreducibly complex and evidence for creation.

**Thou art my God, and I will praise thee: thou art my God, I will exalt thee.**

**~ Psalm 118:28**

In the first few chapters of Genesis, it states that Adam and Eve had three sons – Cain, Abel and Seth. So where did Cain find a wife, and where did everyone else come from? Cain must have married his sister. What? His sister? That can't be; we have laws against brothers and sisters marrying each other because that would produce deformed children.

In Genesis 3:20, we read, "And Adam called his wife's name Eve; because she was the mother of all living." Eve was the first woman, and all the people other than Adam are descendants of Eve. So, Cain would have had to marry his sister. At that time in history, the human gene pool was more perfect than it is today. Siblings could marry each other, as we note with Abraham marrying his half sister (Genesis 20:12). During the time of Moses (approximately 2600 years sfter creation), laws forbidding close relatives marrying were enacted (Leviticus 18-20).

We often think only of those sons mentioned in the Bible: Cain, Abel, and Seth. However, Jewish historian Josephus wrote,"The number of Adam's children, as says the old tradition, was thirty-three sons and twenty-three daughters." How could they have 56 children? In Genesis 5:5, it states that Adam lived for 930 years. With this long life span, having many children would seem logical, especially when God commanded them to "be fruitful and multiply" (Genesis 1:28). Genesis is an accurate historical record, not fantasy.

**And the days of Adam after he had begotten Seth were eight hundred years: and he begat sons and daughters. ~ Genesis 5:4**

In the Garden of Eden, before the Fall, what did mosquitoes eat? Today, the male mosquito drinks nectar from a flower. Only the female drinks blood. The female needs the hemoglobin found in blood to make eggs.  Interestingly, some plants also have hemoglobin.

October 19th

# CHRISTIAN TRUTH INSPIRED BY

Before the Fall, maybe the female mosquito fed on the hemoglobin produced in plants.

Today, more people die from mosquito bites than from any other death caused by insects.  Did mosquitoes bite in the Garden of Eden?  Apparently not because God originally described creation as "very good".  Therefore, there could not have been disease or suffering. God put Adam and Eve in the Garden and said they could eat from any tree except from the tree of the knowledge of good and evil.  Adam and Eve had a choice; they were not created as robots. The primary sin of mankind was, and still is, not trusting and obeying the Word of God.   Adam and Eve chose to eat the fruit of that tree. When they broke the one rule God had given them, everything went wrong. "Cursed is the ground because of you; … both thorns and thistles it shall grow for you …" All of creation was affected … "For we know that the whole creation groans …"  Even the tiny mosquito became cursed.

It is true that Adam and Eve brought the curse of death and disease upon all mankind, but we personally each choose to continue this distrust of God and His Word.  Someday, when Jesus returns, mosquitoes will not harm us.  Until then, let every mosquito bite remind you of how sin entered the world and how all creation groans, waiting for the time of Jesus' return.  At that time, there will be no more crying or pain … or mosquito bites.

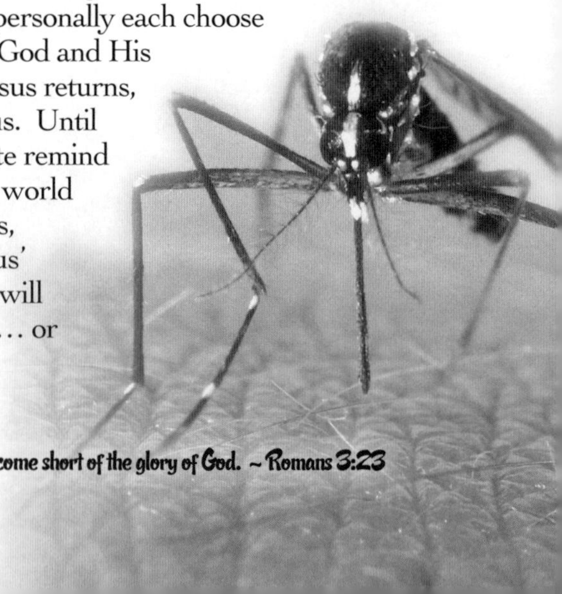

**For all have sinned, and come short of the glory of God. ~ Romans 3:23**

Woodpeckers love to eat bugs that make their homes in trees. To get these bugs, woodpeckers need to have several features. The woodpecker's beak is unlike any other birds; it is so tough that it won't shatter when it hammers a tree hundreds of times a minute. The stiff tail feathers create a tripod with the woodpecker's two short legs. The foot of a woodpecker has two toes in front and two toes in back; whereas, most birds have three toes in front and one in back. This allows the woodpecker to move all around a tree trunk.



When the woodpecker starts to hammer away at the tree, he could easily get a headache, but he doesn't because he has a built-in shock absorber in his head. Special cartilage between his head and beak absorb the pounding. Even with all these features, the woodpecker would still starve to death if he had a normal bird's tongue, that is about the length of the beak. However, the woodpecker's tongue is four times longer than its beak. Its long, slender tongue is used to probe inside insect tunnels in the tree. When it finds an insect, the woodpecker uses the barbs on the tip of the tongue to poke the insect. The tongue is coated with a sticky glue-like substance that glues the insect onto the tongue so it remains secure while drawing to out of the tree. This special glue does not stick to the woodpecker's beak. When the woodpecker is done eating and ready to fly away, where does he store this long tongue? He reels it in and wraps it around the back of his skull.

Why and how would the woodpecker "evolve" all these special features if he could already eat bugs from the ground? God designed the woodpecker from the beginning to be a woodpecker!

**Great is the LORD, and greatly to be praised; and his greatness is unsearchable. One generation shall praise thy works to another, and shall declare thy mighty acts. ~Psalm 145:3-4**

What are the implications of the many ancient records about dinosaurs/dragons? For man to have seen and written about dinosaur-like creatures, they must have survived the flood of Noah and not died out millions of years ago. Here are just a few accounts:

- A Sumerian story (c. 2000 B.C.) tells of a hero named Gilgamesh, who, when he went to cut down cedar trees in a remote forest, encountered a huge, vicious dragon. He slew it, cutting off its head for a trophy.
- The city of Nerluc in France was renamed in honor of the killing of a dragon. This dragon was bigger than an ox and had long, sharp, pointed horns on its head. This could be a description of a triceratops.
- In Europe, during the 1500's, a well-known science book called the *Historia Animalium,* stated that dragons were not extinct but were extremely rare and small.
- An Irish writer (c. 900 A.D.) recorded an encounter with a beast that sounds like a stegosaur. This large beast had a tail with "iron" nails that pointed backwards. Its head was shaped like a horse, and its legs were thick with strong claws.
- An ancient Roman mosaic (2nd century A.D) shows two long necked dragons.
- The emblem on the flag of the country of Wales is a dragon.

October 21st

## BIBLICAL ACCURACY INSPIRED BY

Ancient peoples all over the world have recorded their encounters with these reptile-like creatures that once roamed the Earth. Dragons and dinosaurs have been written about, depicted in art and told about in stories passed down verbally for centuries. Man and dinosaurs did live together from the beginning, having been made on Day six, just as the Bible records.

**And God made the beast of the earth after his kind, and cattle after their kind, and every thing that creepeth upon the earth after his kind: and God saw that it was good. ~ Genesis 1:25**

Opals are said to have formed over millions of years. Yet Len Cram of Australia, who earned his Ph.D. in opal research, found that he could form opals within months. Len was curious and wanted to discover the secret of how opals were formed. His opals are so good that experienced opal miners and scientists in the opal industry cannot distinguish Len's opals from natural opals.

Len's recipe contains an electrolyte, silica, water, some alumina and feldspar. His basic ingredient is a chemical called tetraethyl-ortho-silicate, which is an organic molecule containing silica. The amount of alumina determines the hardness of the opal. Len basically begins with the sandy grit in which natural opals are found, and then he mixes in an electrolyte. Seams of opal then begin to "grow". These seams are identical to what is observed in nature. Interestingly, takes about three months (not millions of years) for the opals to "grow". Rapid opal growth fits the Bible's timescale.



The best explanation for the formation of opals is that during the catastrophic global flood, thick layers of sediments would have been laid down under water, initiating many rapid chemical processes. It did not take millions of years for opals to form – just the right conditions. Len Cram of Australia knows that, and by using that knowledge, he has been able to make millions… by making opals in months, not millions of years! Opals, with their dazzling beauty, declare that the Earth is young.

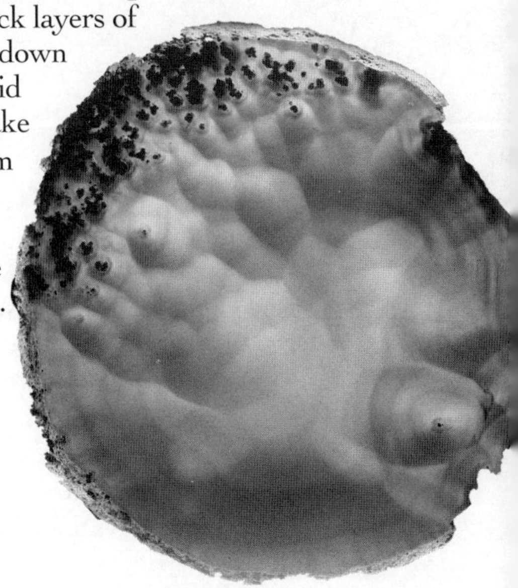

**He putteth forth his hand upon the rock; he overturneth the mountains by the roots. He cutteth out rivers among the rocks; and his eye seeth every precious thing. – Job 28:9-10**

Evolutionists say the moon is 4.5 billion years old. For its size and age the moon then, should have been geologically dead for the past 3 billion years. However, ever since telescopes have been available, there have been reports of bright spots, color changes, hazes, veils and other strange things being seen on the moon. Since these activities are short-lived, they are called Transient Lunar Phenomena (TLP). In the Aristarchus region of the moon, more than 300 of these TLP have occurred.

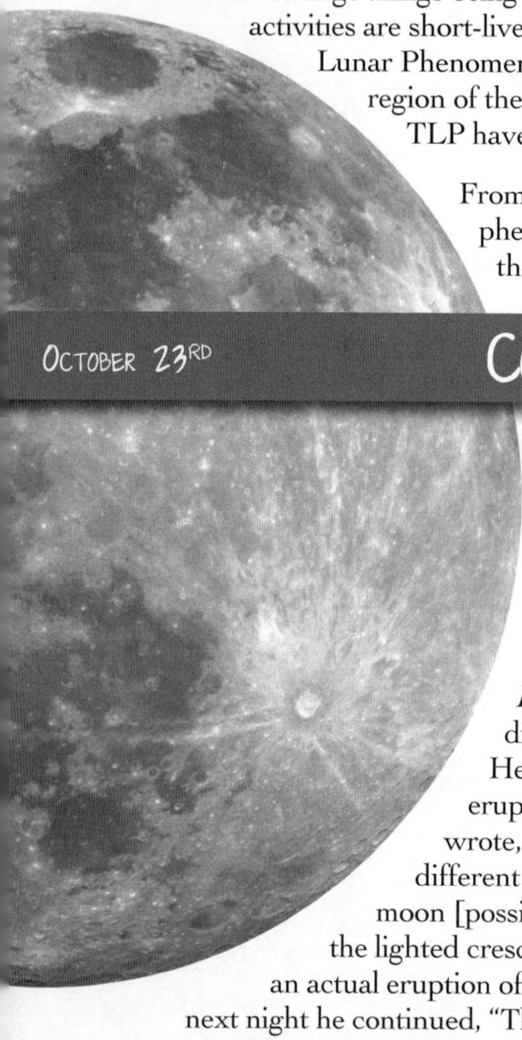

From 1900-1960, many of these phenomena were ignored because the moon was supposedly 4.5



billion years old and therefore dead. But the sightings became so overwhelming that in 1968 NASA published a report titled *Chronological Catalog of Reported Lunar Events*. As far back as 1787, the discoverer of Uranus, William Herschel, recorded volcanic eruptions on the moon. Hershel wrote, "I perceive three volcanoes in different places of the dark side on the moon [possibly referring to an area outside the lighted crescent area]… (the one) shows an actual eruption of fire, or luminous matter." The next night he continued, "The volcano burns with greater violence than last night. I believe the diameter . . . to be about three miles." These observations by Herschel and others point to a young moon.

**And when you look up to the sky and see the sun, the moon and the stars—all the heavenly array ~ do not be enticed into bowing down to them**

**~ Deuteronomy 4:19**

Coral reefs display some of the most spectacular natural color on our planet. Corals are actually animals, not plants. They live mainly in warm, sunlit, well-oxygenated tropical waters. A coral reef is the mass of limestone made as polyps (individual coral) build their skeletons. It was once thought that it took millions of years for a reef to grow. Secular scientists have suggested that the growth rate of reefs is from 0.003 to 0.1 inches per year. However, the discovery of a substantial clump of coral growing on a modern shoe (fewer than four years old) off the Philippines waters suggests this is not the case.

## INSPIRED BY BIOLOGY — OCTOBER 24TH

Ariel Roth of Geoscience Research Institute has examined the way these coral measurements were taken. He found these measurements were taken at the surface that would show lower rates of growth because of exposure to air during low tide, intense sunlight and pounding waves. A much better way of finding the rate of growth would be from depth-soundings. When this was done it revealed a growth rate of 11-16 inches per year. This makes the evolutionary estimate as much as 5,000 times too slow!

The growth rate at the Eniwetok Atoll in the Marshall Islands places the entire atoll at less than 3,500 years old! Scientists have been finding that coral growth rates can nearly double when the temperature is increased 5 degrees Celsius; remember, Eniwetok Atoll sits on an extinct volcanic cone. It does not take millions of years for coral reefs to form, just the right conditions. Even the world's longest coral reef, Australia's Great Barrier Reef, could have formed in the 4,300 years since the Flood.

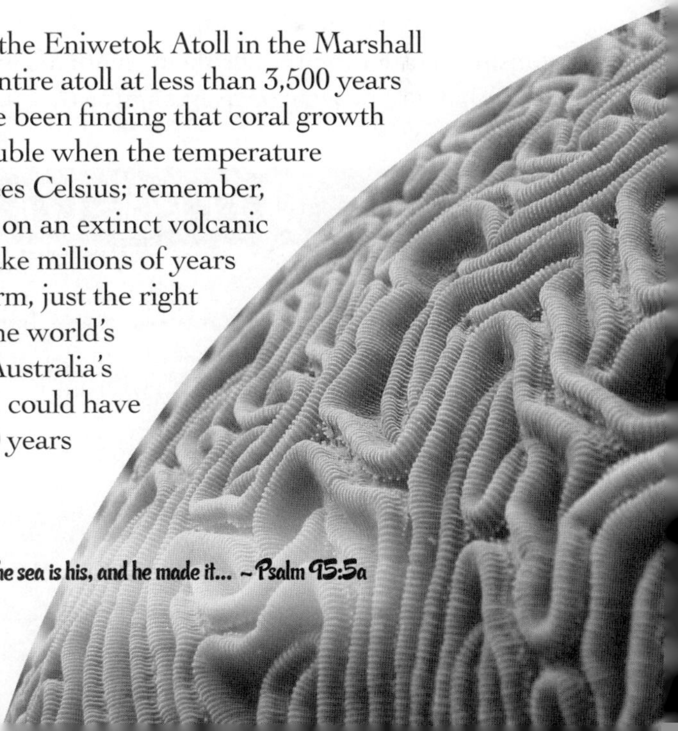

**The sea is his, and he made it... ~ Psalm 95:5a**

The news media often promotes evolutionary discoveries in a way that can best be called exaggerations but, in reality, are simply deceptive. An example is the 2010 announcement that science had succeeded in producing life in the lab from non-living chemicals. This was typical of the media spin on the research:

> "Dr. Craig Venter and his team have built the genome of a bacterium from scratch and incorporated it into a cell to make what they call the world's first synthetic life form."[1]

You have to study the actual research[2] to learn that the single-celled organism they produced was neither made from scratch nor synthetic.

- First, they used DNA from a living cell as a template.
- Second, they used yeast cells to bring DNA fragments together because human science produces DNA fragments one hundred thousand times smaller than needed for life.
- Third, they used a living cell to "proofread" the resulting DNA and correct errors.
- Fourth, the DNA was useless until placed inside another living cell with the thousands of other chemicals and processes needed to keep it alive.

In the end, they had to rely on three different living, single-celled organisms in order to produce their "synthetic, from scratch" life form. What was proven, once again, was that the theoretical law



of abiogenesis is not valid. That theory says that life can arise spontaneously from non-life molecules under proper conditions. In reality, life only comes from pre-existing life!

**And he answered and said unto them, "Have ye not read, that he which made them at the beginning made them male and female"**

**- Matthew 19:4**

It is reported that supply and demand is what causes frequent fluctuations in gasoline prices. This is not a new situation for the world's economy. Consider the economic crisis of the 1700's – the lack of paper. Paper is so common today that it is hard to imagine that there ever was a paper shortage. However, at that time paper was made from cotton and linen rags. These were in short supply because of increased literacy and commerce. English law even required the dead to be buried in wool in order to save the cotton and linen rags for paper making. Crisis? No, opportunity!

An entomologist, Rene Antoine Reaumur, noticed that wasps would chew the fiber of trees and make paper wasp nests. Reaumur said, "They (American wasps) teach us that paper can be made from the fibers of plants … and seem to invite us to try whether we cannot make a fine and good paper from the use of certain woods." The wasps showed scientists what was possible – that the wood cellulose fibers could be separated and remade into paper, providing an almost unlimited supply.

## INSPIRED BY DESIGN — October 26th

We can thank God that He has already provided the solution for any crisis we face. God often places the answer right before our eyes; we just need to observe what He has already made to find the answer to our problems.

**My help cometh from the Lord, which made heaven and earth.**
**~ Psalm 121:2**

Have you considered how precious you are to God? You are, according to the Bible, "the apple of His eye." So what does that mean? When you look into your eyes, do you see an apple? Well, yes, to the writers of the Old Testament, the pupil of the eye did resemble the shape of an apple. They saw an apple (pupil) in your eye. This precious pupil allows sight. It must be protected. So the eye is placed deep within the skull with bones in the forehead and



cheeks to encase it. Meanwhile, the eyebrows and eye lashes catch dust, debris, and sweat. To protect from blindness, eyelids close while sleeping.

The protection continues with the strongest reflex in man—the blinking reflex. If anything gets near the cornea of the eye, blinking happens automatically. This blinking reflex is also the last reflex to exist prior to death. When a nurse or doctor wants to see if a deeply comatose patient has died, they bring a piece of cloth near the cornea of the eye; if they blink, they are still living. If they do not blink, they have probably passed. To be the apple of God's eye is to experience His supreme protection. God protects us reflexively, even unto death. The best example is the death of His Son on the cross. That is how precious you are to God; you are the apple of His eye.

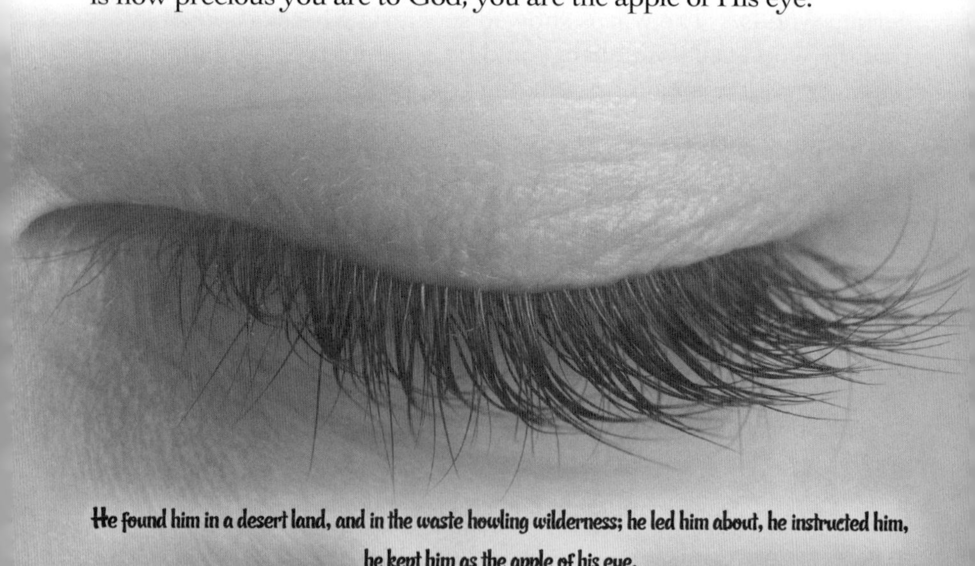

**He found him in a desert land, and in the waste howling wilderness; he led him about, he instructed him, he kept him as the apple of his eye.**

**~ Deuteronomy 32:10**

Lake Missoula formed behind an ice age glacier as it melted. This ancient lake was 1000 feet deep and one-half the size of Lake Michigan - near the present day city of Missoula, Montana. As more and more ice melted at the end of the Ice Age (about 3,600 years ago), it filled this lake and pressure increased. Finally, the ice dam burst, causing the entire lake to empty within a few days. Water rushed at an estimated 60 mph across Washington state, accelerating to more than 80 mph through the narrow Columbian Gorge and then rushed out into the Pacific Ocean.

As these waters careened across Washington, they carved right through the hard lava creating the Channeled Scablands. One of the hundreds of channels dug by this flowing water is now called the Grand Coulee. It is 1,000 feet deep and 50 miles long and was formed by flowing water in less than two days! Throughout the



Channeled Scablands are numerous large basalt boulders, some as large as houses. Also strewn along the flood path are large erratic boulders that apparently floated along in the flood within icebergs from the shattered ice dam. One boulder, transported to Willamette Valley of Western Oregon, originally weighed over 30,000 pounds (until tourists started removing pieces!)

If this one relatively tiny flood could so radically carve the landscape – imagine the whole world flooded during Noah's time.

*For, lo, he that formeth the mountains, and createth the wind...*
*The Lord, The God of hosts, is his name.*
*~Amos 4:13*

The hummingbird burns lots of energy as it darts from flower to flower with his wings beating some 80 times per second. The hummingbird uses up so much energy that if it goes for more than two hours without eating, it will die. Yet the exhausted hummingbird sleeps eight hours every night, so why doesn't it die? During the day, when the hummingbird heart beats 10 times a second; while at night when the hummingbird sleeps, the heart slows down to less than one beat per second. Also, during the day the hummingbird's temperature is 100°F; while at night, it drops to 50-60°F. This enormous drop in temperature would kill most other



warm-blooded animals. That would be like our heart slowing to less than 10 beats a minute and our bodies cooling to 48°F while we slept – we'd be rushed to the nearest hospital's intensive care unit (or the morgue)!

By dropping its temperature and slowing its heartbeat, the hummingbird is able to get a good eight hour sleep without needing food. God has provided the hummingbird with a remarkable and unique metabolism.

The hummingbird has to be the most amazing flying machine ever envisioned. It can beat is wings up to 80 beats per second; fly backwards, forwards, and sideways with ease; fly up to 50 miles per hour; hover in place; refuel in flight; and migrate thousands of miles.

To accomplish such feats, it needs lots of fuel – nectar from flowers. The hummingbird gets this nectar using its long, thin beak as it hovers over the flower. Hummingbirds do not sip up nectar like using a straw. The bird has a special tongue that is curled up at the edges to form two troughs. The tongue moves in and out of the bill



at an unbelievable rate of 13 times per second. Each time its tongue retracts into its beak, nectar is squeezed down the throat. When finished, the tongue, which can stick out farther than the bill, is pulled back and wrapped around the skull!

The hummingbird's ability to hover is enabled by a unique swivel "shoulder" joint that allows it to produce a figure eight with its wings. To fly backward, the hummingbird tilts its wings ever so slightly, so the air is forced forward. All this happens so quickly that the wings are a blur to the human eye. No other bird has this feature, and there is no fossil transition showing how it could have developed by evolutionary processes.

Now imagine a half-evolved hummingbird having the ability to hover but with the beak of a sparrow – it would not be able to get nectar. Or imagine having the long beak but no ability to hover – it would have to stop and perch. Flowers do not have perches. God designed the hummingbird from the beginning with its tongue, long beak, and hovering ability so it could get nectar, its fuel for life.

**I meditate on all your works and consider what your hands have done.**

**- Psalm 139:5**

What keeps rain falling around the planet to bring water to the plants and animals? It is the Earth's water cycle. But where does the water come from when most of the Earth's water is contained in salt-saturated oceans, which is harmful to most life?

Four thousand years ago, the Bible declared that water is "drawn up" out of oceans to come down again as rain. Water indeed evaporates, mostly from the oceans, leaving the salt behind. The water vapor is transported by the winds to precipitate elsewhere. Heat from the sun drives this hydrological system. Heat warms up the Earth, making billions of gallons of water vanish from puddles, rivers, lakes and oceans. Evaporation into water vapor is a slow process because of water's large heat capacity. It takes a lot of heat to turn a small amount of liquid water into vapor. Without this high heat content requirement, lakes would soon disappear. Yet, the amount of energy coming onto our planet from the sun and the enormous reservoir of water in our oceans provide abundant evaporation to water the entire Earth even though the process is slow.

This water vapor then rises into the atmosphere where it cools and condenses into liquid drops. These droplets combine and grow until they are too heavy to remain in the air and then fall to the Earth as precipitation (rain or snow). Precipitation is hard



to predict and impossible to control, but rain and snow water the Earth. There is no similar water cycle on any other planet. Our water cycle is truly unique. God displays His power through what He has made.

*For he maketh small the drops of water: they pour down rain according to the vapour thereof: Which the clouds do drop and distil upon man abundantly.*

*~ Job 36:27-28*

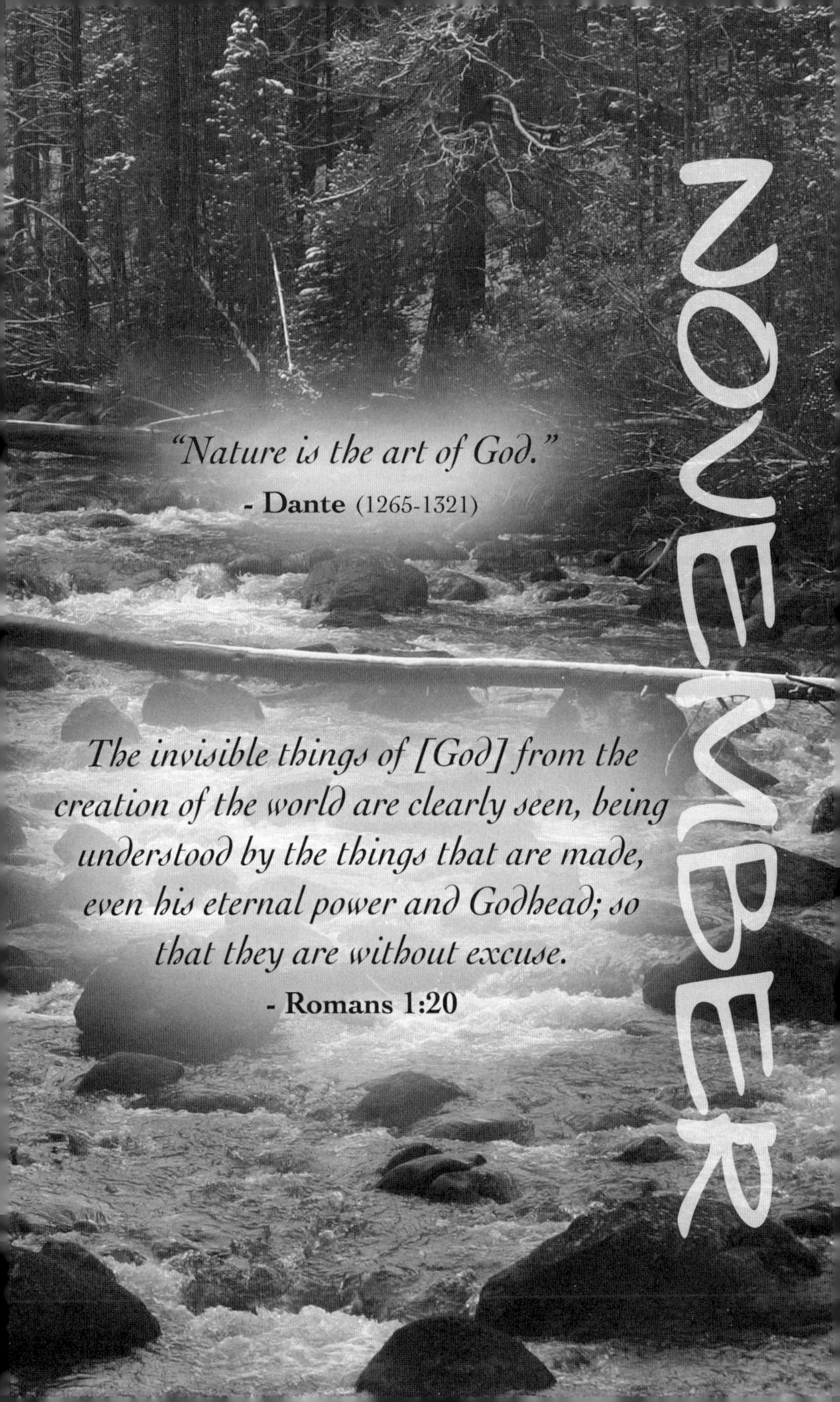
NOVEMBER
"Nature is the art of God."
- Dante (1265-1321)
The invisible things of [God] from the creation of the world are clearly seen, being understood by the things that are made, even his eternal power and Godhead; so that they are without excuse.
- Romans 1:20

A person's worldview colors his interpretation of every aspect of life. The Biblical worldview acknowledges that God created the entire universe (the basis for understanding physics and cosmology), that there has been a recent creation of very different forms of life



(the basis for understanding biology), that sin and death exist as a result of mankind's actions (the basis for understanding human psychology), the reality of a worldwide flood (the basis for understanding geology), and that humanity dispersed of across the globe after the Tower of Babel (the basis for understanding world history).

A non-biblical worldview must come to a completely different conclusion in all of these areas of study. Inevitably, these explanations leave God out and involve enormous periods of time. A biblical worldview acknowledges that God made everything but also accepts responsibility as a steward over this creation. Therefore creation is to be used, utilized, and wisely preserved, but there is also an understanding that it was made for our enjoyment. The idea that global warming or climate change is some sort of crisis to be prevented shows an incredible ignorance of true world history. The Earth's climate has been in a constant state of fluctuation and reverberation since the world-wide flood, which totally resurfaced the planet only about 4500 years ago. Thus any recent moderate changes are totally insignificant when viewed through the larger biblical perspective.

A non-biblical worldview sees any environmental change as a crisis of enormous importance. This is because this worldview sees the planet as being billions of years old and mankind having only been present for a tiny fraction of this time. Therefore, any effect we are having on creation is considered detrimental and evil. In effect, the planet becomes the god to be worshipped and preserved. This is the real driving force behind the furor over climate change, global warming, the green movement, and minimizing mankind's "carbon footprint".

**[Man] changed the truth of God into a lie, and worshipped and served the creature more than the Creator, who is blessed for ever. Amen.**

**- Romans 1:25**

Have you considered where all the languages came from? About 7000 distinct languages exist in the world today. Linguists have

## INSPIRED BY HISTORY

NOVEMBER 2ND

tried to find the original language from which all other languages originated. They have utterly failed.

The Indo-European language family contains 45% of the different forms used on the planet. This includes English, Spanish, Russian and 446 other languages. The next most widely spoken language family is the Sino-Tibetan group which includes Chinese. It is estimated to have about 403 distinct languages – approximately 22% of the world uses languages from this group. Besides these two families, 92 other distinct families exist. No one has shown how language families as different as Indo-European and Sino-Tibetan could have derived from a common source. The same is true of the other 92 families. Linguists have simply hit a wall in explaining the many distinct language families throughout the world.

The biblical account of the Tower of Babel is the only explanation that fits the data. The Bible says God confounded the language at Babel to separate the people. God miraculously changed the common language into many languages. Where did the many distinct languages come from? From God Himself.

**And the whole earth was of one language, and of one speech. So the LORD scattered them abroad from thence upon the face of all the earth: and they left off to build the city. ~ Genesis 11:1,8**

The grape vine requires severe annual pruning to remain fruitful. All growth from the previous year, except for one or two nodes, is pruned – cut off. If this process is not followed, the vine grows wildly in every direction producing much foliage but little fruit. When properly pruned, the grape vine produces an abundance of flavorful fruit and juice, which play a significant role in feeding much of the world's population. Grapes are the most widely grown fruit in the world.

Jesus used pruning as an analogy for helping us understand God's process of making us more useful for His kingdom. God is not as concerned with our "happiness" as our "holiness". He puts far more emphasis on our spiritual depth than our material wealth. Our Creator is not a genie called upon to produce instant healing on demand, but He is an all-knowing physician who will allow physical challenges as a method for refining us into more fruitful followers. We would be wise to look at the physical, material, and spiritual challenges as God's "pruning process". Whatever we are going through is being allowed by God. If we remain strongly attached to Him as our source of strength no matter how difficult the trial, the ultimate outcome will be lives transformed for the glory of God.

**I am the true vine, and my Father is the husbandman. Every branch in me that beareth not fruit he taketh away: and every branch that beareth fruit, he purgeth it, that it may bring forth more fruit.**

**~ John 15:1-2**

A fifty pound alligator snapping turtle is a voracious eating machine capable of consuming everything in sight and stripping a pond of fish and other creatures. Unlike some fish which adapt to colder temperatures and continue eating through the winter, the snapping turtle becomes dormant as water temperatures fall, and it discontinues eating for months.

The alligator snapping turtle is also designed with a totally different respiration system than other creatures. A turtle's lungs are confined within a rigid shell that cannot expand by using muscle



contractions like other creatures. Thus, its respiratory muscles are designed to force air out of the lungs, and air pressure refills the lungs when the muscles relax. This is the opposite of human lungs, which are empty when relaxed and filled when the muscles contract. Alligator snapping turtles can also draw oxygen directly from water into their blood stream using a special chamber called the cloaca. By utilizing both lungs and cloaca, they can stay submerged for days once the water temperature drops below 50°F.

The intricate design of each feature of this turtle makes it absolutely apparent that this creature is the result of brilliant design rather than random step-at-a-time changes. From its ability to process air by different methods to shutting down its metabolism (thereby allowing it to survive through the winter), the turtle makes the genius of its Creator apparent.

**Unto thee, O God, do we give thanks, unto thee do we give thanks: for that thy name is near thy wondrous works declare.**

**- Psalm 75:1**

Rock layers seldom show erosion between the different strata. For the most part, they have essentially flat, knife-edged boundaries. If evolution was true, we should see erosional features between the rock layers. Just look across landscapes today – there are hills and valleys, rivers and streams, all cutting ruts through any landscape. Yet, the Grand Canyon reveals huge regional rock layers without these up-and-down erosional features.

The "bathtub" ring at the top of the canyon (as the Park rangers like to call it) is the Coconino Sandstone. Right below it is the Hermit Shale. Between these two layers is a flat, featureless boundary where no erosion has occurred. If millions of years passed between the deposition of these sediment layers, where are the erosional features? Throughout the canyon we find the same knife-edge, flat, featureless boundaries between the layers. We find this same geological feature around the world, flat. These knife-edge



boundaries between the rock layers indicate that the sediments were **NOT** laid down slowly over millions of years.

The year-long Genesis Flood would have laid down the rock layers rapidly and continuously with no time for erosion to take place. The lack of erosional features between the rock layers shows rapid deposition, that perfectly fits the Genesis Flood model for geology…but this observation creates enormous difficulties for the billions of years evolutionary model.

*The earth is the Lord's, and the fulness thereof; the world, and they that dwell therein. For he hath founded it upon the seas, and established it upon the floods.*
*- Psalm 24:1,2*

How did the "bathtub" ring, i.e. the Coconino sandstone, form within the Grand Canyon strata? This sandstone has a thickness of 315 feet and covers a visible area within the Grand Canyon of 200,000 square miles and extends across much of the U.S. Within this sandstone is an easily noticed feature that geologists refer to as cross-bedding. Cross-bedding is sand laid down at an angle or incline from the horizon. Cross-bedding can be the result of either water or wind. Sand grains are moved across the dune and tumble down the face of the dune. Research has shown that the desert dunes' cross-beds have a slope angle of 33-34 degrees, while underwater sand waves create cross-beds with a slope angle of less than 30 degrees. The Grand Canyon's Coconino sandstone was made by water-deposited sand. Ocean currents, not violent desert winds, moved the sand and rapidly produced these cross-beds.



Sand mounds produced in the laboratory were found to correspond to water depth, i.e. as the water depth increased, so did the sand dune height. Sand dune heights were generally found to be about one-fifth of the water depth. Water velocities were also found to affect sand dune characteristics. Using this data, the depth and velocity of the water responsible for creating the Coconino sandstone cross-beds were calculated. Cross-beds with a height of 30 feet imply deposition by sand waves at least 60 feet high, water 300 feet deep, and a current velocity of 2 - 3.75 mph. These current velocities would have to be sustained in one direction for days. The closest thing to these conditions observed today is a tsunami.

How appropriate for the Grand Canyon National Park authorities to call the Cocanino Sandstone the "Bathtub Ring" because water from Noah's flood was what created this sediment layer.

**The voice of the LORD is upon the waters: the God of glory thundereth: the LORD is upon many waters. ~ Psalm 29:3**

Have you had your appendix out? Many doctors over the last 150 years have based their thinking on Charles Darwin's evolutionary theory and were trained to believe the appendix was a useless, vestigial organ. Therefore, it was no longer needed. Modern medical research has shown that Charles Darwin was wrong.

The appendix is not something left over from our evolutionary



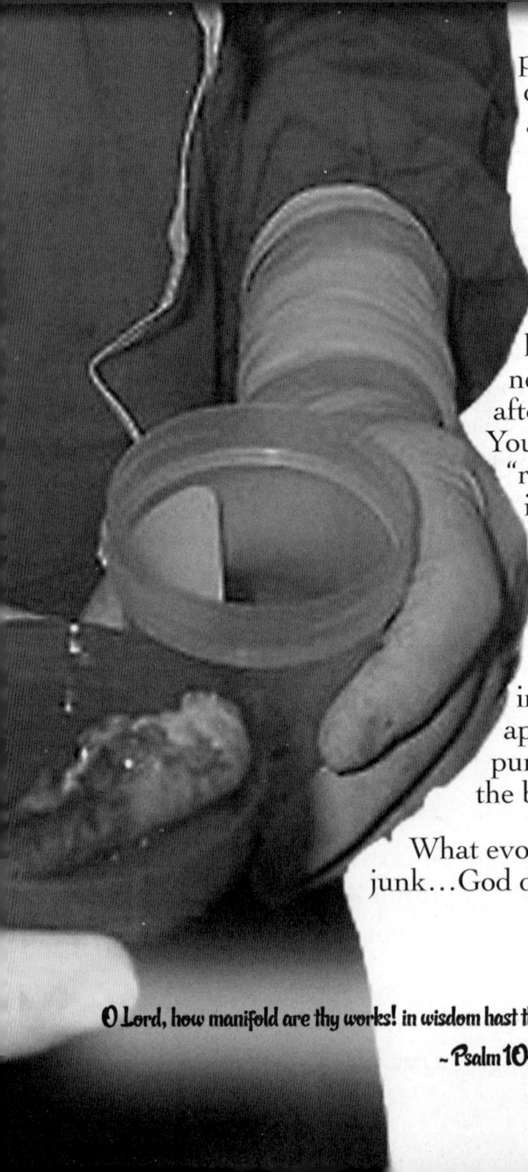

past. Scientists have recently discovered that that the appendix is in fact a very, very useful feature of the human body. Researchers at Duke University Medical Center have found that the appendix "is a safe haven where good bacteria could hang out until they were needed to repopulate the gut after a nasty case of diarrhea".[1] You could say that your appendix "reboots" the good bacteria into your intestines. Careful study does not show that the appendix emerged/evolved from animals lacking one, nor does it show any slight, continuous modifications in the fossil record[2]. The appendix was planned with a purpose and created by God in the beginning.

What evolutionists label as useless junk…God designed for our benefit.

**O Lord, how manifold are thy works! in wisdom hast thou made them all: the earth is full of thy riches.**

**~ Psalm 104:24**

## INSPIRED BY COSMOLOGY

NOVEMBER 8TH

The moon creates the tides that are essential for healthy oceans. As the water rises and falls in their daily rhythms, the shorelines are continually scrubbed. Tides stir the oceans, preventing the water from becoming stagnant and unhealthy. If there were no stirring of the oceans, they would become one big stagnant pond. Our oceans contain vast amounts of floating plankton, kelp forests and grasses. In fact, there is more plant life by weight in the oceans than on the land. Plants take in our waste product, carbon dioxide, while releasing oxygen. If there were no tides, then the ocean waters would move very little, plant life would decrease – thus making less oxygen – and our atmosphere would rapidly deteriorate. We then would have insufficient oxygen for life. Our very breath depends on the lunar tides.

**The day is yours, and yours also the night; you established the sun and moon. ~ Psalm 74:16**

Have you ever tried to stealthfully approach a grazing deer? Chances are it bolted long before you got close. The eyes of the deer are uniquely designed to actually see nearby and distant objects simultaneously. This allows a buck to graze while at the same time keep a keen eye out for predators. Amazing! Notice how a deer's eyes are set high and wide – this allows a deer to see almost in a complete circle. It is incredibly hard to sneak up on a deer!

Did you know that deer also have special night vision "goggles"? Because a deer is most active in the early morning or late evening when there is little light, he has been designed with night vision.



Special reflective cells (tapetum lucidum) are situated at the rear of its eyes. First, the light enters the eye and is absorbed. Then the light which was not absorbed is reflected back toward the retina a second time - allowing it to be absorbed. We see these reflective "night vision goggles" at work when our headlights expose the eyes of a deer – they seem to glow back at us. As if this ability still wasn't enough, deer can also see in the ultraviolet portion of the spectrum – light which is invisible to humans.

How did an eye create the lens needed to see both near and far simultaneously? How did an eye develop a reflective layer? How did the deer's brain learn to "see" in the ultraviolet spectrum? Atheists just wave the magic word "evolution" and pretend that explains it all. There is a better answer – these abilities come from the Designer's hand.

As the deer pants for the water brooks, so pants my soul for You, O God.

- Psalm 42:1

What happens when a scientist uses a biblically based model for his science? The end of scientific advancement? Science dragged back into the dark ages? Not quite.

Dr. Russ Humphreys, a Ph.D. physicist and biblical creationist, used biblical assumptions and produced the best model available for predicting the magnetic fields of the planets. His model used physics to estimate the initial strength of the magnetic field of each planet in our solar system at the moment of creation if it were made "out of water". Then his model computed the planet's expected magnetic field strength if this creation were 6,000 years ago.

Using this model, Dr. Humphreys predicted the magnetic field strength of the planets Uranus and Neptune before the Voyager spacecraft measured them. He predicted that Mercury would still have a magnetic field when all other models predicted it should be long gone. Dr. Humphreys also predicted that Mars would have remnant (permanent) magnetism. Remnant magnetism takes place in rocks that cool and solidify in the presence of an external magnetic field. The moon has also been found to have such remnant magnetism. In every single prediction based on biblical



starting assumptions of a recent creation, Humphreys' model was amazingly accurate. If the planets were really billions of years old, their magnetic fields should be extremely weak, but they are not. These magnetic fields confirm that God created the solar system about 6,000 years ago.

**And the earth was without form, and void; and darkness was upon the face of the deep. And the Spirit of God moved upon the face of the waters.**

**- Genesis 1:2**

Many people believe that Genesis 1 and 2 contradict each other, in part because the animals seem to be created in a different order. In Genesis 2:19, the King James Bible states, "out of the ground the LORD God formed every beast of the field and every fowl of the air". It seems that land beasts and birds were created between Adam and Eve. Compare that with Genesis 1:23-25 which indicates that Adam and Eve were both created after the beasts and birds. What's going on?

In the original Hebrew language, the word translated "formed" in Genesis 2:19 would have been understood to mean "had formed". Now read Genesis 2:19 with this understanding, "Now the Lord God had formed out of the ground …" The apparent disagreement with Genesis 1 completely disappears – the animals had already been formed.

Other people say there are two completely different "myths" of creation – Genesis chapters 1 and 2. In reality chapter 1 is creation from God's perspective (the big overview), while chapter 2 is creation from man's perspective. Chapter 2 doesn't mention the creation of the heavens, Earth, atmosphere, seas, land, sun, stars, moon, or sea creatures. Chapter 2 only mentions things directly relevant to Adam and Eve and their life in the garden.

## November 11th BIBLICAL ACCURACY INSPIRED BY

Jesus himself recognized both Genesis 1 and 2 as historical truth. When Jesus was speaking about the subject of marriage in Matthew 19:4-5, he referred to Genesis 1:1, Genesis 1:27, and Genesis 2:24. Jesus Christ did not regard them as two separate, contradictory accounts. God's word is true and reliable from the very beginning.

**And he answered and said unto them, "Have ye not read, that he which made them at the beginning made them male and female", And said, "For this cause shall a man leave father and mother, and shall cleave to his wife: and they twain shall be one flesh?" - Matthew 19:4-5**

The clown fish makes its home in the deadly sea anemone. The sea anemone is basically a stomach and mouth surrounded by stinging poisonous arms. When a fish brushes up against the sea anemone, thousands of microscopic stinging cells fire and stun the fish. Then the arms slowly move, working the fish toward the anemone's mouth which is in the center of the arms. This den of death is where the clownfish calls home!

The clown fish lives among these stinging arms without dying. How? The clown fish is coated with the same mucus that covers

the sea anemone. How did the clown fish get this protective mucus? Did he steal a sample and then learn to produce it himself? How did the clown fish know he needed this protective mucus if he wanted to live with the sea anemone? How many clown fish tried different types of mucus before they knew which one would protect them? Evolution would have us believe that the right kind of mucus on the clown fish happened by accident and chance. God designed the clown fish from the beginning with this protective mucus, so it could live with the sea anemone.

**Behold, how good and how pleasant it is for brethren to dwell together in unity! ~ Psalm 133:1**

Ancient people around the world referred to large reptile-like creatures as dragons. If you examined the descriptions of these creatures, they fit very nicely with the characteristics of dinosaurs. But are dragons really dinosaurs?

## NOVEMBER 13TH CHRISTIAN TRUTH INSPIRED BY

The word "dinosaur" did not exist until 1842, when Sir Richard Owen coined the word dinosaur. Dinosaur means "fearfully great lizard or reptile." Prior to 1842, people called them dragons. The Chinese are renowned for their dragon stories with depictions of dragons prominent on Chinese pottery, carvings, and embroidery. Some old Chinese books even tell of a family that kept dragons and raised the babies. In those ancient times, it is said that Chinese Kings used dragons for pulling the royal chariots on special occasions!

When Alexander the Great (c.330 BC) and his soldiers were marching into India, they found hissing dragons in caves. It was reported that people from India worshipped these huge reptiles. England tells the story of St. George who "slew a dragon". The list could go on and on; worldwide, there are hundreds of dragon accounts. These historical descriptions of dragons fit the characteristics of dinosaurs and a biblical worldview.

**Leviathan...smoke pours from his nostils...flames dart from his mouth...**
**- Job 40:1,20,21**

Birds do not sweat because they do not have sweat glands. But they must get hot from all that flying, so how do they stay cool? A bird is cooled from the inside out. Birds are constantly active and generate a lot of heat. Therefore, they need a specially designed cooling system. It all begins at the bird's lungs. Connected to the lungs are seven to nine air sacs that are connected to the major hollow bones. When a bird breathes, cool fresh air moves into the lungs and air sacs and then into the hollow wing and leg bones. As air flows into the hollow bones, the excessive heat produced from flying is quickly removed. This is the ultimate in an air-cooled engine!

The respiratory system of the bird does more than bring in oxygen for it to breathe; it also keeps it cool. What if the bird did not have this type of cooling system? Bird flight is a vigorous activity that produces lots of heat. If the heat could not be removed quickly and efficiently, the bird would overheat and die. God created birds with this special air cooling system.

*Gavest thou the goodly wings unto the peacocks? or wings and feathers unto the ostrich?*

*- Job 39:13*

Our eyes are very delicate organs. Daily, they are exposed to dust, dirt and debris. Tears are our main defense against this constant bombardment of contaminants. Human tears are made up of more than 90% water and contain a special enzyme that destroys bacteria. Imagine if this enzyme was not present – our eyes would be one big continual infection. With this infection would come pain; remember the last time you got something in your eye? So every time you blink, stop and consider how your eye is washed with the original "antibacterial soap" – the one created by God.

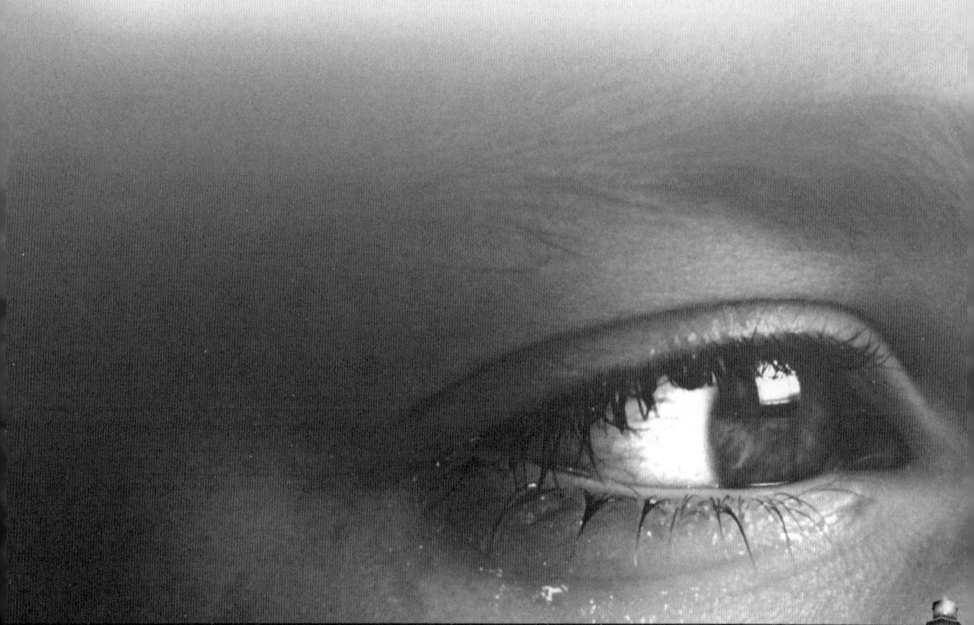



**And the eye cannot say unto the hand, I have no need of thee...there should be no schism in the body...**
**~1 Corinthians 12:21,25**

The camel is designed to survive and thrive in the desert. Deserts are blistering hot during the day and freezing cold at night. While humans have to keep a very precise body temperature (98.6°F) to avoid serious harm, camels can vary their temperature without ill effects. A camel can allow its body temperature to rise as high as 106°F so that it does not feel so hot. When it is cooler at night, the camel can reduce its temperature, so it does not feel so cold.

A camel has an especially slim body. This narrow shape is ideal when the hot midday sun shines – less area is exposed. The camel's



especially long legs enable the camel to be higher off the ground and away from the heat radiating from the hot sand. The camel's urinary system is amazing in that it concentrates the urine in order to save water. Camels have a special drinking capacity. When water is readily available, a camel can drink 26 gallons in only 10 minutes.

Even if deserts did not exist on the original pre-flood paradise-like earth, the amazing information and variability for the camel's desert survival features had to have been "programmed in" – they could not have arisen by chance. The camel needed all these features from the moment deserts existed in order to survive and thrive in the hot sandy environment.

***...And thou shalt remember all the way which the Lord thy God... brought thee forth water out of the rock of flint; Who fed thee in the wilderness...***

***~ Deuteronomy 8:2,15,16***

Have you considered the origin of the moon? Many careers and billions of dollars have been spent working to solve this question. There are still many theories because none of them adequately explain the existence of our moon.

- **The Collision Theory or the Big Whack:** A massive object hits Earth. The impact hurled rock out into space. The rock fragments began to circle around our planet and turned into our moon. *Problem:* How does a ring of debris form a moon?
- **Fission theory:** The Earth spun so fast that a chunk broke off. The moon came from the Pacific Ocean. *Problem:* The Earth could never have spun fast enough to throw a moon



into orbit, and moon rocks were found to differ from Earth's material.
- **Capture Theory:** The moon was wandering through space and Earth captured it. *Problem:* How do you slow down a wandering moon and insert it in Earth's orbit? It would either crash into the surface or be thrown outward with great speed, like in the game crack-the-whip.
- **Condensation, Nebular, or Accretion Theory:** The initial gas clouds and dust compressed, creating our Earth and moon. *Problem:* Material in a vacuum doesn't just fall together into a big lump.

So, what is the origin of the moon? The best answer is both the simplest and the most profound. God made the moon.

**It shall be established for ever as the moon,**
**and as a faithful witness in heaven. ~ Psalm 89:37**

The lotus plant grows in a muddy, watery environment very much like the water lily. Yet, its leaves always seem spotlessly clean. How does a lotus plant keep itself clean even in muddy waters?

The leaves of the lotus are large and quickly shed water. When the leaves were examined, they were found to be waxy with microscopic bumps. These countless tiny bumps were the key to its cleanliness. They cause the rain drops to break the surface tension forming droplets that then rolled off taking dirt and grime with them. A company copied the lotus leaf and now sells a coating that is applied to furniture and leather that "repels" water, dust, and grime. Roof tiles are being designed like the lotus leaf – with each rainfall, the roof is cleaned without detergent. Some new paints coming on the market are also applying this "lotus effect"; with each rainfall, the house is cleaned without detergent. There is even an umbrella based on the lotus leaf that is instantly dry when shaken.

When we are faced with a problem like how to remove dirt and grime without detergent, it is good to look to see how God has solved this problem. Then we can just copy the design.

Enormous cavernous rooms, large enough to hold Super Bowl sized stadiums, can be found within the cave systems of Carlsbad Caverns in New Mexico. It is commonly taught that these caves were created over millions of years as a very weak carbonic acid solution (formed from water leaching through calcium carbonate layers) slowly dissolved the limestone layers to form the massive cave system. The rate at which this process is happening today would require millions of years to explain the current Carlsbad cave system with its enormous cavernous rooms.

What is overlooked is another mechanism that explains the formation of these caves with a far more rapid process. Adjacent to the cave system is a volcanic rich area that could easily have leached sulfur dioxide into water flowing through the Carlsbad cave system. This would have formed a much stronger sulfuric



acid capable of dissolving limestone much faster than the weak carbonic acid process. Furthermore, the product which results from a reaction of sulfuric acid and limestone is a mineral known as gypsum, and there is a vast deposit of ultra pure gypsum, known as the White Sands National Monument, immediately adjacent to the Carlsbad cave area. This 'smoking gun' is hardly a coincidence.

There is no reason to believe colossal cave systems, such as Carlsbad Caverns, could not have easily formed within the time since Noah's flood some 4500 years ago.

**Thou visitest the earth, and waterest it: thou greatly enrichest it... Thou waterest the ridges thereof abundantly: thou settlest the furrows thereof... ~ Psalm 65:9,10**

Have you considered that man has always been highly intelligent?



- Sacsahuamn, an ancient city built by the Incas near Cuzco, Peru, contains a massive wall of irregularly shaped blocks of stone. These blocks fit together so well that a piece of paper cannot fit between the joints after centuries of earthquakes. One incredibly-sized rock is larger than a five-story building and weighs 40 million pounds! How did the builders move and fit these irregular stones together?
- In Baalbeck, Lebanon are the ruins of a Roman temple. It stands on a single foundation stone weighing an estimated 4 million pounds. This foundation stone was set in place prior to the time of the Romans. How were these ancient people able to quarry and move this stone?
- The "Gateway of the Sun" at Tiahuanaco, Bolivia contains a single cut stone of volcanic rock weighing an estimated 200,000 pounds. How was it transported?
- Stonehenge of England: These massive stones were quarried from over 75 miles away and then moved into place. Some of the stones weigh an estimated 200,000 pounds. Stonehenge appears to be an extremely accurate astronomical calculator. How did the builders move these massively heavy slabs to their precise positions?

In Genesis 4:21-22, we find early mankind working with bronze and iron. By the time of the Flood (some 1,600 years later), man would have reached a high level of technology. After the Flood and the Tower of Babel, man spread out over the Earth, taking their knowledge with them. Man has been intelligent from the beginning.

**... now nothing will be restrained from them, which they have imagined to do.**

**~ Genesis 11:6**

One of the major problems for evolution, apparent even to Darwin when he first published his theory, was the lack of support from the fossil record. To this day, there is no adequate explanation for why all of the basic body types (phyla) suddenly appear at the lowest layer of sedimentary rocks (Cambrian). For instance, how did some single-celled ancestor turn into a starfish and leave no transitions in between? What is the transition between single cells and clams or jellyfish or coral or squids or lobsters or fish? All of these vastly different creatures appear suddenly, fully formed and fully functional, with no transitions.

The late Harvard paleontologist, Dr. Stephen Gould, supposedly solved this mystery with a theory called "punctuated equilibrium". Your children will be required to accept this as an explanation for why the fossil record does not show transitions. According to Gould, we do not see creatures in the process of turning into completely different creatures (half formed features) because the changes are happening TOO SLOWLY to see significant change happening. Yet, according to Gould, when change did happen, it happened



TOO FAST (or in too isolated of a location) to have been captured in the rock record, i.e. a "punctuated change to the equilibrium".

Thus in an incredible leap of faith, evolution has become the only area of science where the LACK OF EVIDENCE is promoted as the evidence for believing this theory!

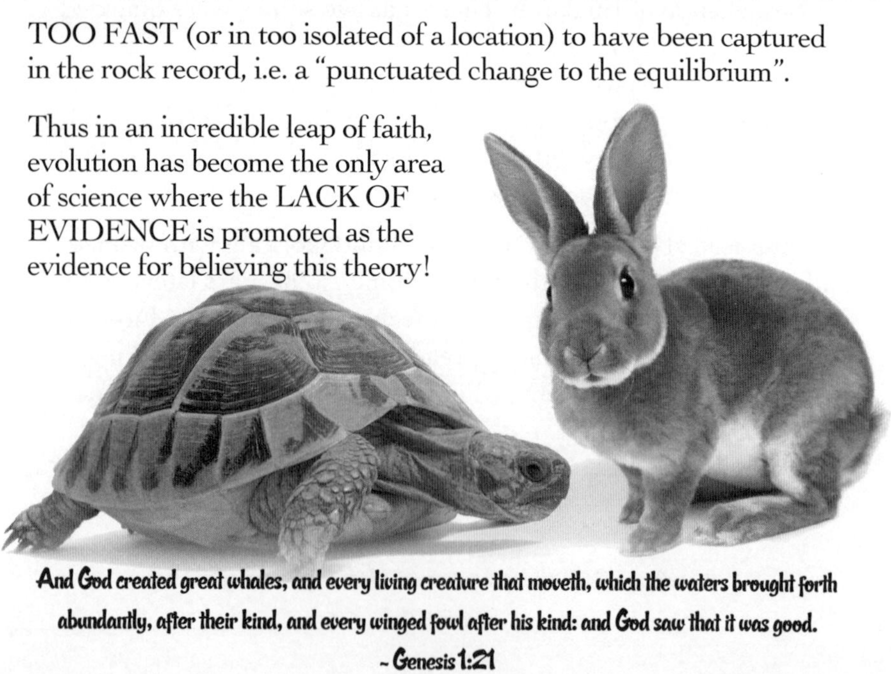

**And God created great whales, and every living creature that moveth, which the waters brought forth abundantly, after their kind, and every winged fowl after his kind: and God saw that it was good.**

**- Genesis 1:21**

A Weddell seal can dive 2000 feet down to enormous water pressures and then return to the surface quickly without coming to any harm. If human divers tried this, they could die of the bends (caused by dissolved nitrogen forming bubbles within the blood). To prevent experiencing the bends, divers ascend slowly with periodic stops. If they ascend too quickly, they have to spend time in a decompression chamber, or death results. Why don't seals get the bends and die?

## INSPIRED BY BIOLOGY



Before submerging, seals completely empty their lungs of air. They still have oxygen in their bodies, but it is distributed throughout in their blood and muscles, rather than held as a gas in their lungs. In fact, a seal's body can store ten times more oxygen than the human body. As seals dive, their heartbeat slows, thereby using less oxygen. Also, oxygen is directed away from organs such as the liver and directed to essential organs like the brain, which need more oxygen.

Seals are able to dive effortlessly. How did seals know how to store extra oxygen in their blood and muscles? How many seals died from surfacing too quickly before their bodies developed the intricate mechanisms needed to allow them to dive deeply and surface quickly? Dead seals would have had a little trouble warning others not to dive so deep! The fact is, God designed seals from the beginning to have the ability to dive deep without experiencing problems.

***They that go down to the sea in ships, that do business in great waters; these see the works of the Lord, and his wonders in the deep.***

***- Psalm 107: 23-24***

Stanley Miller's famous experiment in 1953 attempted to show that life began when lightning passed through a particular atmosphere and made chemicals called amino acids. Amino acids are the building blocks of proteins, and proteins are required by all living cells. Miller did show that some amino acids could be made in this way, but it's quite another thing to get them to build a living cell! Just as a few concrete blocks in a parking lot will never turn themselves into a complex building, the existence of a few amino acids can never explain the origin of a living cell.

These amino acids supposedly linked together to form proteins (like beads on a necklace). These proteins then somehow would have to form the first DNA that went on to form the first living cell. Generations of schoolchildren have been taught this fairytale as if it were a fact of history. Now even many atheists are bailing out and admitting life could not have possibly formed in this way. Why?

(1) Amino acids in water do not concentrate themselves; they disperse.

(2) Amino acids need to be in the pure form in order to make proteins. Contaminants in ocean water would have stopped protein formation.

(3) Under natural conditions, pure amino acids will not form proteins.

(4) Living things use only left-handed amino acids, yet Miller-type experiments always result in a useless 50:50 mixture of right and left-hand amino acids.



What Miller's experiment actually showed was that life could not possibly have formed in this way. The complex organization of life requires an intelligent Creator! It is by observing creation that God makes His awesome power and creativity apparent to everyone.

**Because that which may be known of God is manifest in them; for God hath shewed it unto them.**

**- Romans 1:19**

Migration: The movement of birds to a warmer or colder place. Even though birds are small, they have incredible sensors within their bodies that allow them to navigate by using the sun, or constellations, or landmarks, or the Earth's magnetic field.

## INSPIRED BY CHRISTIAN TRUTH — NOVEMBER 24TH

Migration is even mentioned in the Bible in a surprising way. Jeremiah 8:6-7 illustrates repentance in terms of bird migration.

Migratory birds that fly to warmer climates in fall, always return the following spring. We also need to be continually returning to God with a repentant heart. Perhaps God placed migration within nature as a frequent example to us of how we need to return to the warmth of God's loving forgiveness. The next time you see birds in migration, think about any sins in your life from which you need to turn away in order to fly back towards the Lord.

...No man repented him of his wickedness, saying, What have I done? every one turned to his course, as the horse rusheth into the battle. Yea, the stork in the heaven knoweth her appointed times; and the turtle and the crane and the swallow observe the time of their coming; but my people know not the judgment of the Lord. ~ Jeremiah 8:6-7

Elephants can weigh up to 11,000 pounds. That is a lot of weight their poor feet to support! Elephants spend a great deal of time along river banks and watering holes. They have been known to sink into the mud. Have you ever stepped into mud and tried to pull your foot loose? It is a very difficult task, often resulting in the loss of your shoe! As you try to pull your foot loose, a vacuum actually develops around your foot preventing it from coming smoothly out of the mud. So why don't elephants, with larger feet and far more weight, get stuck in the mud?



It is the design of their feet that makes all the difference. Elephants walk on their tiptoes with thick fat surrounding their bones. When an elephant lifts his legs, the diameter of his legs actually gets smaller – making it easier to remove them from the mud. It is hard for an elephant to get stuck in the mud! Evolution teaches that such features happened by accidental chance changes. Tell this to the first elephant born without this feature. These earlier "less-evolved" elephants would be constantly getting stuck in the mud, waiting for the first lion to come along and have them for lunch. That would be a quick end to elephant evolution.

When God designed the elephant, He thought of everything.

Come and see the works of God:
he is terrible in his doing toward the children of men.
~ Psalm 66:5

Are you eating a dinosaur for Thanksgiving? Many evolutionists say that that the dinosaurs never went extinct; they only evolved into birds. So, are you eating a dinosaur for Thanksgiving? Let's examine the evidence. Reptilian lungs are very different than bird lungs. Reptilian lungs are similar to human lungs in that air flows in during inhalation and out during exhalation. Bird lungs, however, have air flowing straight through the lungs into air sacs located throughout the body. These are two totally different systems.

Reptilian scales are completely different than feathers. Evolutionists have long argued that feathers evolved from reptile scales; however, this is storytelling, not science. Feathers grow out of follicles. This tubular down-growth of epidermis

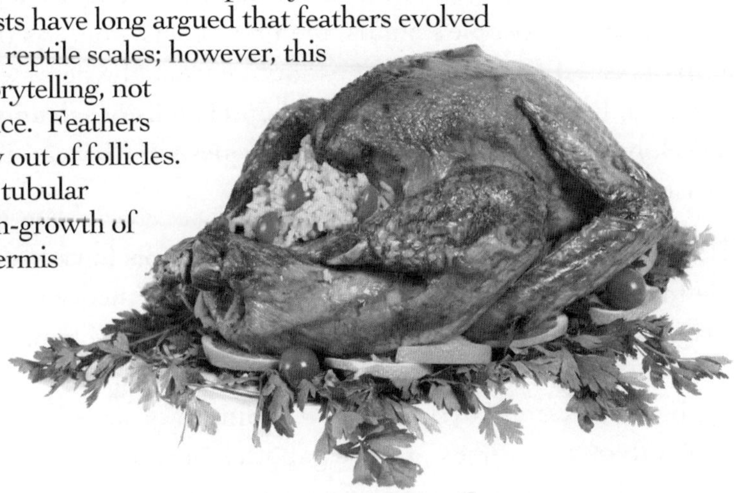

extends deeply into the skin (in the case of primary feathers, all the way down to the bone). At the base of this follicle, the feather is produced. Reptile scales are nothing but folds in the skin, like fabric folded over on itself – again, two totally different designs.



So, you are not eating a dinosaur for Thanksgiving. God, the Master Designer, created each according to its kind; a bird is a bird, and a reptile is a reptile.

**They, and every beast after his kind, and all the cattle after their kind, and every creeping thing that creepeth upon the earth after his kind, and every fowl after his kind, every bird of every sort.**

**- Genesis 7:14**

Genesis Chapter One is one of the most God-centered chapters in the Bible. In only thirty-one verses, God is mentioned 32 times. If we add the use of the personal pronoun, He is mentioned 43 times. Thus, on the very first page of Scripture, we are brought into the presence of God. How Satan hates this! He has waged a full-out attack to discredit the first chapter of the Bible!

If Genesis chapter one can be transformed to mere mythology or story-telling in people's minds, then Satan has won. As John Phillips asks, "If the Holy Spirit cannot be trusted when He tells of creation, how can He be trusted when He tells of salvation? If the Holy Spirit cannot be trusted in Genesis 1, how can He be trusted in John 3:16?"

In the last several centuries, wide-spread attempts have been made to discredit Genesis 1 as literal truth. One theory is the day-age theory; the days of creation are not literal 24 hour days but epochs or ages or vast periods of time. The gap theory has a gap between Genesis 1:1 and 1:2 – adding enormous ages between these two verses. In this gap, they insert all the "time" demanded by atheists and evolutionary geologists. However,



to the Hebrew reading this passage, the days of creation were obviously six regular 24-hour days. God's word is true from the very beginning. We can trust God at His word; our very salvation depends on it!

Rain forests of the world usually receive 12 feet or more of rain each year. That much rain would wash out most of the soil's nutrients. So how do rain forest trees receive the necessary nitrate fertilizer needed for their continued growth? The answer - bats.

Many huge rain forest trees are hollow inside. Some varieties of fruit bats find this an inviting place to sleep during the day. Fruit bats eat fruit and then transport these raw materials to the trees. Inside the tree, the bats' rich droppings, guano, provide

the trees with one of the richest known sources of nitrates. The trees provide a protective place for the bats, and the bats provide fertilizer for the tree. The bats feed the trees! This mutual arrangement makes possible the tropical rain forests, which provide homes to millions of creatures. What an ingenious method our Creator has devised to provide for His many creatures.

**... both small and great beasts....that thou mayest give them their meat in due season. That thou givest them they gather: thou openest thine hand, they are filled with good. - Psalm 104:25,27-28**

Have you noticed that the universe is expanding? Since we are positioned here on Earth, it is impossible to feel, but astronomers assure us that this is the case. In the book of Isaiah, God clearly told us that he stretched out the universe, so it should be no surprise that astronomers have been able to confirm this fact. This prediction must have seemed very strange when first written. The universe does not appear to be expanding. If you went out on a starry night, it would look about the same size as the previous night. Even when we examine ancient star maps, they appear essentially identical to today's night sky.

## November 29th — Cosmology Inspired By

It was not until the 1920s that astronomers concluded that virtually all galaxy clusters appeared to be moving away from the other clusters – indicating that the entire universe is indeed expanding. You can observe this same effect with a balloon and a marker. Take a marker and put dots onto a balloon. Each dot represents a galaxy. Inflate the balloon. You will notice that the dots move farther away from each other. If the entire universe were being stretched out, the galaxies would all be moving away from each other, and that is exactly what they appear to be doing. Yet, only at a very special location, the center of the balloon, would all dots be moving away from you. This Earth seems to be located at this very special location where the entire universe is being stretched away in every direction – we are located essentially at the very center of the universe! The Bible wrote of an expanding universe some 2,700 years ago, and we are just beginning to understand all of the implications.

**"(God) stretches out the heavens like a curtain,**
**and spreads them out like a tent to dwell in".**
**- Isaiah 40:22**

Our bodies produce tears in the lacrimal glands just above each eye; the tears flow across our eyes and drain into the nasal cavity. Our body makes tears continually; we only notice them when we cry, and then the excessive tears overflow the ducts and roll down our cheeks. Scientists have discovered that the chemical composition of tears that are cried while peeling onions differs from tears expressed in response to strong emotions.



Onion tears are the result of irritation by sulfur-based chemicals from the onion. These tears protect your eye from possible chemical harm. When tested, emotional tears were found to contain measurable quantities of the following chemicals:

- manganese (a chemical depressant)
- leucine-enkephalin (chemical which helps control pain)
- adrenocorticotrophic hormone (a chemical produced by bodies under stress).

When you cry, these "toxic" chemicals are excreted and released – thus, we actually feel better after a good cry. Interestingly, humans are the only creatures that cry for emotional reasons. God made us with the ability to express and feel strong emotions. He also gave us the capacity to deal with those emotions. One way to release the build-up of these emotional chemicals is to have a good cry. Perhaps it was your mother who first said, "Why don't you have a good long cry, and you will feel better."

God designed our body marvelously, and even the shedding of tears is a gift from the Designer.

**For the Lamb which is in the midst of the throne shall feed them, and shall lead them unto living fountains of waters: and God shall wipe away all tears from their eyes. ~ Revelation 7:17**

# DECEMBER

*"The truth is incontrovertible,*
*malice may attack it,*
*ignorance may deride it,*
*but in the end; there it is."*

**- Winston Churchhill** (1874-1965)
British Statesman, Prime Minister

*Jesus saith unto him,*
*I am the way, the truth, and the life:*
*no man cometh unto the Father, but by me.*

**- John 14:6**

Why does beauty exist? Beauty can display itself in many ways: as patterns, brightness, curves, and variety. Even though beauty cannot be quantified, beauty is real. Beauty is so important that manufacturers spend great amounts of effort and money to make products appealing to the eye. Beauty can be seen in the classical Greek columns. All that was needed to hold up ancient temples were plain cylinders, yet the Greeks chose to embellish their columns with grooves and carvings. The Greeks wanted beautiful columns.

We see the same thing in nature, for example, the peacock's tail. A recent study from the University of Tokyo found no evidence that peahens chose mates based on the tail feathers. These researchers had wanted to confirm Darwin's idea that peacocks choose mates based on the beauty of the tails. They were surprised that females mated with "poor quality" males as often as with those having "high quality, flashy" feathers. In others words, the actual observations of science contradicted what was predicted by evolutionary teaching. So why isn't the evolutionary theory discarded? Because evolution is a religious dogma, not a scientific theory.

What is the purpose of the peacock's gorgeous tail? Perhaps it is there for man and His Creator to enjoy. The tail of the peacock is

breathtakingly beautiful. When we go to the zoo, this is one of the creatures we want to see. Why? Because the tail shows exquisite beauty. The Bible teaches us that God deliberately put beauty into the world. When we see beauty in this world, we should be reminded of the beauty Maker – God Himself.

**And out of the ground made the LORD God to grow every tree that is pleasant to the sight, and good for food; the tree of life also in the midst of the garden, and the tree of knowledge of good and evil.**

**~ Genesis 2:9**

During Darwin's time, only the cell wall, cytoplasm, and nucleus could be seen. It was not until the advent of advanced microscopes that we could peer deeply into a cell. What scientists found amazed them! They found a micro-city.

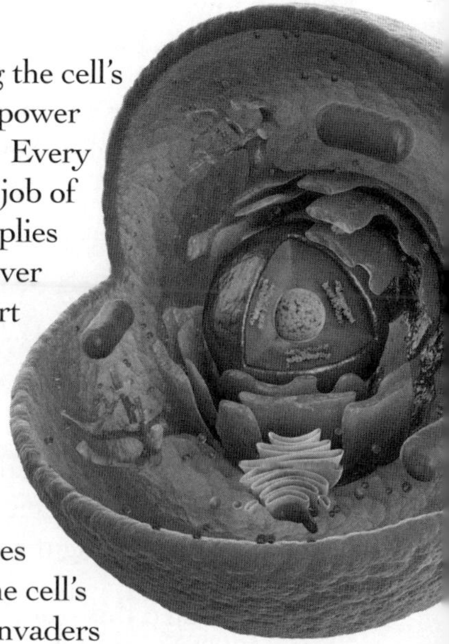

The nucleus is like the city hall, directing the cell's activities. The mitochondria is the cell's power plant, giving the cell its energy to work. Every city needs grocery stores, and that is the job of the Golgi bodies. Golgi bodies store supplies of chemicals that the cell makes. Whenever proteins or fats are needed in another part of the cell, the Golgi body wraps them up and sends them to where they are needed. The endoplasmic reticulum transports things within the cell like a mailman. It also acts like a garbage collector, picking up waste, so the cell does not become polluted. The lysomes are the cell's police force, protecting it by destroying invaders (like bacteria). They also send trash out through the city wall (the cell membrane).

Darwin never knew all the activity that was going on within a microscopic cell. It really is like a miniature city abuzz with activity. Does a city build itself by accident and chance? Just like



a city takes planning and organizing, so too, our not-so-simple cells needed a planner and organizer. That planner and organizer was God who ensured that each part was present and working from the beginning.

**Now unto him that is able to do exceeding abundantly above all that we ask or think, according to the power that worketh in us ~ Ephesians 3:20**

Have you considered the yellow-tailed goatfish? It is a white fish with a yellow tail. Schools of goatfish hang around reefs where the angelfish live because the goatfish love to eat the angelfish for lunch. However, the goatfish have a problem; they are bothered when their gills become infested with parasites.

When the parasites get to the point of driving the goatfish crazy, they see red. Actually, they turn red – literally! Once red, the goatfish swim to the reef where the angelfish live. When the angelfish see a red goatfish, instead of swimming away in terror,



they swim over to the goatfish and nibble off the parasites. When all the parasites have been removed, the goatfish turns white again and swims off, leaving the angelfish in momentary peace.

Who negotiated this truce? Why would an angelfish help the enemy? Why would the angelfish trust the enemy that it would not be eaten? Why doesn't the goatfish have a nice angelfish meal once it has been cleaned? Did this happen by accident and chance, or was it designed? Evolutionists wave a magic wand and say all this just "evolved" because it benefits both species – but that explains nothing. God could and does create such complex relationships.

**Let the heaven and earth praise him, the seas, and everything that moveth therein.**

**~Psalm 69:34**

One of the greatest mysteries of all human history is the coming of God to dwell with mankind. The Creator of the entire universe entered his creation as a participant in the human drama – rather than remaining detached from his creation as an observer. The



all-powerful mastermind of time also decided to change human destiny by coming as a babe born in an obscure animal stall, to an unknown teenage girl in a tiny unknown town, and announce his birth to lowly shepherds in a nearby field. Angels announced to these shepherds that there would be a "sign", that this babe would be wrapped in "swaddling clothes".

To understand why a baby wrapped in swaddling clothes would be "a sign", you must understand the culture of that day. Travelers often took long strips of cloth with them on journeys because the fatality rate was so high and travel so dangerous. These strips, called "swaddling clothes", were used to wrap their bodies for transport to a suitable burial place should they die in transit. No one in that culture would wrap a newborn child in clothing meant for burial, unless no other option was possible.

Thus, even the first cloth to touch His body was used by the Lord of the universe to prophetically announce the reason for His coming – to die for our sins – as signified by being wrapped in swaddling clothes at the very moment of birth.

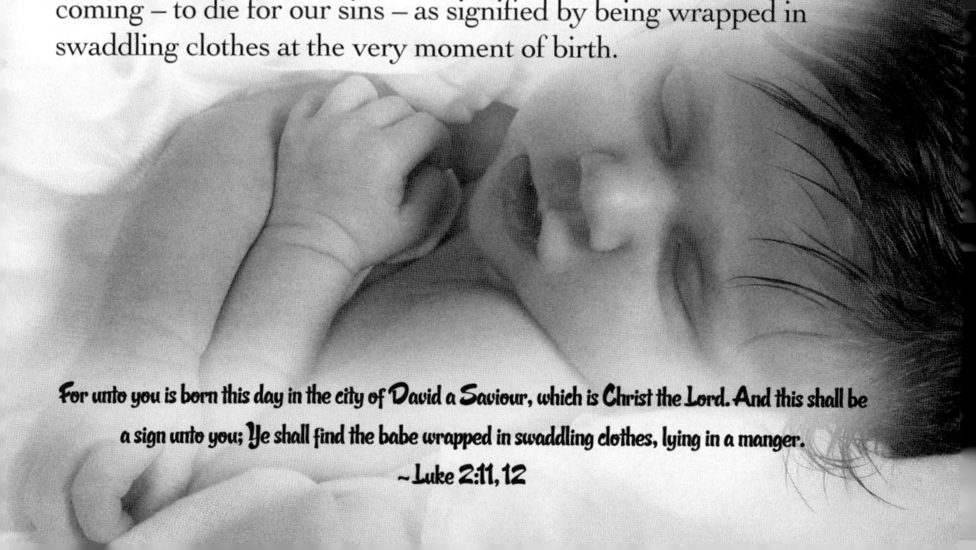

**For unto you is born this day in the city of David a Saviour, which is Christ the Lord. And this shall be a sign unto you; Ye shall find the babe wrapped in swaddling clothes, lying in a manger.**

**~Luke 2:11,12**

According to evolution, over millions of years, the size of the human brain has increased. Evolutionists believe this increase in brain size produced an increase in intelligence. For example, books on evolution show Homo erectus having a brain size of about 1000cc – which is smaller than today's average 1550cc human brain. They imply that this is proof of human evolution.

However, human brain size varies enormously. Human brain volume can range from 700cc to 2200cc. Does a larger skull size mean the person is more intelligent? If so, this is proof that men (with an average brain size of 1600cc) are smarter than women (with an average brain size of 1500cc). Obviously, there is a problem with this assumption. Intelligence has no relationship to brain size.

Consider the computer. Do we say that the modern computers, which are smaller than the computers of the past, are less powerful because they are smaller? No, we pride ourselves in making



smaller and smaller computers with immense processing power. If someone lined up the computers made from 1990 to 2010 and ranked their degree of technological advancement based on their size, they would be incorrect. In a similar way, it is erroneous to conclude that man's intelligence is based on skull size.

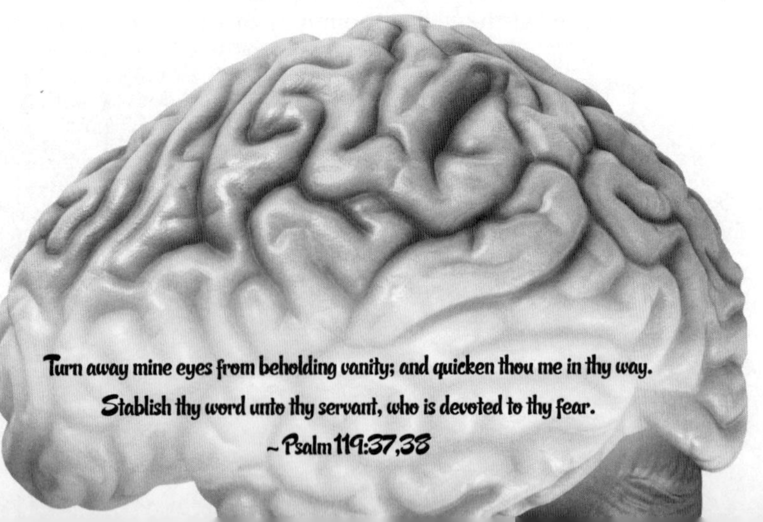

Turn away mine eyes from beholding vanity; and quicken thou me in thy way. Stablish thy word unto thy servant, who is devoted to thy fear.
~ Psalm 119:37,38

Where do pterosaurs, "winged-lizards" (or flying reptiles), fit into history? Evolutionists say these flying reptiles and man did not live together. Yet, according to the Bible pterosaurs must have been created on Day 5, while man was made on Day 6. When the Flood came, Noah apparently took some of these creatures, along with every other created "kind" of land dwelling creature, on board the Ark.

Many cultures from around the world speak of pterosaurs, but they use a different name. In Isaiah 14:29 and 30:6, we find Isaiah mentioning "fiery flying serpents". Or maybe you have heard of the "legend" of the thunderbird. It was a large creature with a bony crest, fierce claws on its wings, and a long, sharp beak. When it flew, its wings thundered. It was a fearful creature to behold! Stories and descriptions of thunderbirds appear in the Indian cultures from Alaska to South America.

Ancient Indian artwork also depicts these flying creatures. For example, one of the Ica Burial stones of Peru displays a carving of a pterosaur. In one of Utah's canyons, along with other ancient Indian pictographs, a flying creature with a crest on its head looking very much like a pterosaur is portrayed. Herodotus (460 BC) writes of flying reptiles in ancient Egypt and Arabia. Even today, there have been reports of these flying creatures in remote



parts of the world (Papua, New Guinea and the jungles of Africa).

Evolution needs to explain away all of this evidence in order to maintain the belief in millions of years separating mankind and other creatures in the fossil layers. Evolution is simply a story attempting to deny the Bible and leave God out. Therefore, it must ignore widespread evidence for the co-existence of man with pterosaurs. Man and flying reptiles did live together.

**Give instruction to a wise man, and he will be yet wiser: teach a just man, and he will increase in learning. - Proverbs 9:9**

The aurora borealis appears as a shimmering band or waving curtain of greens and reds that periodically appear in the dark, northern evening sky. I can remember lying in the back yard in total awe one summer evening for 30 minutes as God painted this magnificent, constantly moving, wondrous light show in the evening sky.

This beauty is a reminder of God's protection. The aurora is radiation made visible. It all begins on the sun's surface where protons and electrons are boiled off the sun's surface. These dangerous, high-speed particles race through space reaching Earth in a matter of hours. Instead of hitting the Earth directly, the particles are swept to the Polar Regions by the Earth's magnetic field. This radiation hits the upper atmosphere causing the air to glow with



energy. Without this magnetic field, we would not survive! God placed this special field around us to protect us from these dangerous incoming particles.

Scientists have discovered a correlation between sun spot activity and auroras. The greater the number of sunspots, the greater the number of auroras. Sun spot activity runs on an 11-year cycle, so 2012 and 2023 should be optimum for observing the aurora borealis. If you ever have an opportunity to see these shimmering bands of green and red – praise God for more than just the beauty you are witnessing – praise Him for the protection He provides.

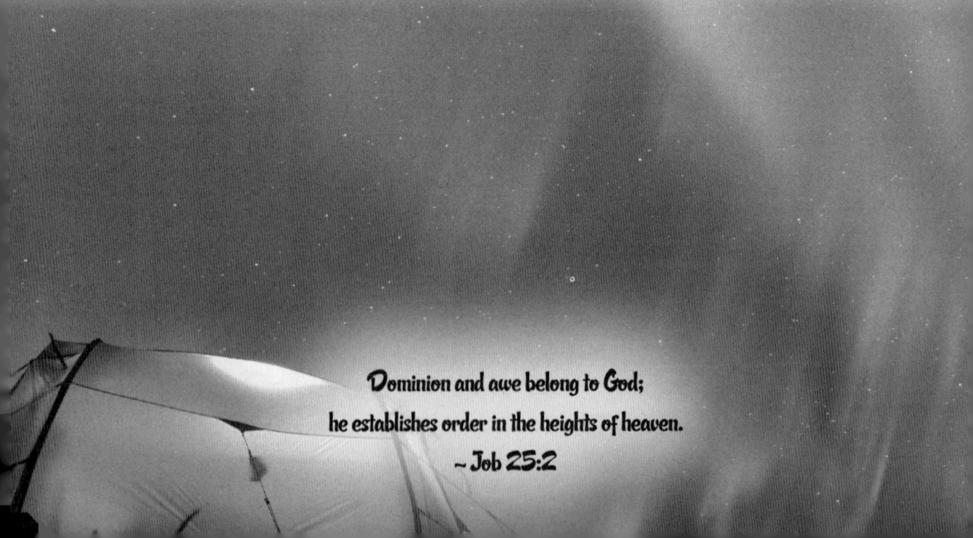

**Dominion and awe belong to God;**
**he establishes order in the heights of heaven.**
**~ Job 25:2**

Have you frozen a frog lately? In the middle of winter, the wood frog shows no sign of life. It has no heartbeat, no breathing and no circulating blood. Yet when winter is over and the ice melts, this frog returns to life as if it had never been frozen. Freeze a human and he doesn't wake up! So how does a frog do this? It goes on a "sugar high".

Before the frog freezes, its blood sugar level (glucose) reaches a very high level; in fact, it reaches an extreme concentration. This excessive level of glucose keeps water from leaving cells and prevents shrinkage. The elevated level of glucose also reduces the



liquid's freezing temperature – allowing only a very small amount of the inner body liquid to turn to ice in the cold. In addition, the glucose feeds the cells through the winter when no other source of nutrition is available. It even prevents metabolic reactions like urea synthesis from taking place. Normally, an extremely high glucose level would kill a frog, so how does this elevated amount of glucose come about just before the wood frog freezes?

As soon as ice appears on the wood frog's skin, a message travels to the frog's liver to convert much of its stored glycogen into glucose. Five minutes after the message is received, the sugar level in the blood starts to increase. Who could believe that this complex set of processes all just happened by accident? God, in His wisdom, even programmed into the DNA code of wood frogs a method for freezing frogs without killing them!

**But I have trusted in thy mercy; my heart shall rejoice in thy salvation. I will sing unto the LORD, because he hath dealt bountifully with me.**

**~ Psalm 13:5,6**

Think about the thick layers of sandstone around the Earth. Let's take a look specifically at Zion National Park in Utah. It consists of rock made from a 2,200 foot thick layer of sand called the Navajo Sandstone. Within this Navajo Sandstone are many layers of sloping mounds or waves of sand. The "waves" indicate that water moved the sand. Where did the sand come from? After surveying 15,000 North American localities and taking more than half a million geological measurements the American Geological Survey has determined that water had to have carried the sand 1250 miles across the continent from the Appalchian Mountains of New York to Utah.

Evolutionists suggest that some large river ran across the continent carrying its sand load over millions of years. But how could water flow consistently for millions of years and carry only pure quartz sand? How could it dump its load in an area of one and a half million square miles? Why didn't it leave sand anywhere in between?



Energetic water currents of a global flood could have transported such huge volumes of sediment right across the North American continent to deposit the sand in such thick layers. The flood of Genesis 7-8 describes exactly such an event. We find thick layers of sandstone not just at Zion National Park but around the world. Only a worldwide Flood adequately explains such geological evidence.

Darwin's three greatest challenges were explaining how life could have developed from non-life, explaining where all the information



originated to create the enormous variety of life, and explaining the presence of enormous gaps in the fossil record (the fossil record shows no process of gradual transformation). All three major objections remain after 150 years of scientific investigation!

To overcome the second problem (information origination), Darwin proposed the idea of pangenesis. We don't hear much about this today because the careful work of Darwin's contemporaries proved his idea to be nonsense. Pangenesis is the idea that every part of an organism's body emits some sort of "information containing particle", that transfers changes from one generation to the next. Darwin referred to this vague information transferring mechanism in various ways: gemmules, granules, particles, atoms, and even cells. Yet, he never produced a shred of experimental evidence as to their existence.

Darwin's half-cousin, Francis Galton, ran a series of experiments in which he transferred the blood of black rabbits into silver-gray rabbits (along with a control group) to see if the off-spring of the purebred silver-gray rabbits would be gray, black, or in between. He bred a total of 124 offspring in 21 generations without producing a single mongrel rabbit. This, and a myriad of other experiments, definitively showed that the information needed to explain the diversity of life was created within the original animal type, not acquired by breeding or through environmental influences.

Essentially, everything Darwin believed about evolution has proven to be incorrect. Amazingly, he is still considered one of the greatest scientists who ever lived.

**O Lord, how great are thy works! and thy thoughts are very deep. A brutish man knoweth not; neither doth a fool understand this.**

**- Psalm 92:5,6**

Your knee has an incredibly intricate design which is often only fully appreciated when it stops working properly. Consider what happens as you're watching a football game and a player goes down. He is holding his knee; he seems to be in excruciating pain. What's happening with his knee? He has probably torn one of the ligaments, in particular the ACL, anterior cruciate ligament.



Did you notice the name of this ligament? It has the word cruciate in it. The word cruciate comes from the Latin word meaning "cross". And indeed, the two ligaments at the knee joint form a cross; the ACL (anterior cruciate ligament) crosses over the PCL (posterior cruciate ligament). When the ACL is torn, it causes excruciating pain. So here in the midst of a football game, the message of Jesus' crucifixion can be told—his excruciating death on a cross for our sins. We are called to use every opportunity to proclaim His message. Who would have thought it could happen at a football game with a torn ACL?

Pilate saith unto them, What shall I do then with Jesus which is called Christ? They all say unto him, Let him be crucified. And the governor said, Why, what evil hath he done? But they cried out the more, saying, Let him be crucified. ~ Matthew 27:22-23

Ram Mountain in Alberta, Canada is world famous for bighorn sheep rams with large horns. Hunters from around the world are willing to pay large sums of money to pursue these prized animals. However, the horn size is in decline. Many hunters are coming home empty-handed, with no sheep meeting the minimum regulation size. Researchers have documented the decline in horn size over the past 30 years as sheep with larger horns are preferentially killed. Are the sheep evolving into small creatures?

## INSPIRED BY BIOLOGY

DECEMBER 12TH

Evolution requires the addition of new functions and features to explain the diversity of life upon our planet. What is happening with these sheep is the removal and elimination of the genes for large horns and bodies. When these genes are removed from the ram populations, what is left are genes for smaller horns and smaller bodies. This is not evolution but the selective elimination of genes for large horns and bodies. The Ram Mountain bighorn sheep population has lost genetic information. This is the very opposite of evolutionary advancement.

It seems that once the genes for largeness have been removed, they are gone forever. At the time of creation, God had put into His critters a vast gene pool. Over time we have eliminated part of that gene pool by selective harvesting. What is taking place within the Alberta bighorn sheep population is a narrowing of the gene pool through artificial selection – yet the sheep are still sheep. They are not turning into goats, horses, or cows.

**Know ye that the LORD he is God: it is he that hath made us, and not we ourselves; we are his people, and the sheep of his pasture. - Psalm 100:3**

People have been fascinated with the moon since the beginning of time. Where did it come from? How big and how far away is it? How old is it? Who would have thought that sending mirrors to the moon would help provide an answer! By bouncing laser light off these mirrors, scientists have been able to measure the distance to the moon with great precision. What they found also has amazing implications for the age of the Earth.

We now know that the moon is moving away from the Earth about two inches each year. The moon moving away from the Earth little by little is called lunar recession. If the Earth is 6,000 years old and the lunar recession has always been happening, then the moon has moved about 1,000 feet away from the Earth. This is no big deal.

But evolutionists teach that the moon is four billion years old. Working backwards at two inches per year, the moon would have been causing impossibly high tides and related problems on the Earth within only 10% of that much time. In less than 50% of the assumed age of the moon it would have been touching the Earth's surface! Evolutionists face a problem with the age of the moon – it simply cannot be four billion years old. The moon's recession is evidence that the Earth/moon were recently created.

Quartzite boulders provide powerful testimony to the global flood. Quartzite rock was once sedimentary sandstone but became a hard metamorphic rock under tremendous heat and pressure. Billions of rounded quartzite boulders and cobbles are found scattered throughout the northwestern United States and western Canada. Where did they come from? How did they become so widely scattered?

The source of quartzite rock is near the Continental Divide in Montana, Idaho, and British Columbia. Yet, we find quartzite rocks scattered eastward and westward up to 600 miles from their source. Many of these hard quartzite rocks have percussion marks that indicate collisions during transport within violently moving waters. The speed of current needed in order to carry boulders over 600 miles into Saskatchewan and North Dakota has now



been calculated. To transport and suspend a rock 6 inches in diameter would require currents of at least 65 mph and water depths of 200 feet. These rates are almost incomprehensible to modern geologists stuck in uniformitarian thinking. Modern flash floods seldom exceed 20 mph yet do enormous erosional damage. The billions of quartzite boulders distributed 300-600 miles from their source are powerful evidence for the watery catastrophe of the Genesis Flood.

**And the waters prevailed upon the earth a hundred and fifty days.**

**~ Genesis 7:24**

One of the main problems for animals living at high altitudes is the decrease in air pressure. Thus many species have hemoglobin molecules that are specially adapted to increase oxygen affinity. The bar-headed goose is one such species that lives around Nepal and Tibetan lakes at altitudes from 12,000-20,000 feet above sea level. It has even been observed flying over the summit of Mt. Everest, 29,000 feet above sea level!

The hemoglobin molecules of the bar-headed goose have been sequenced and compared to close relatives that do not live at high elevations. It was found that there are four changes in the hemoglobin



molecules. The main mutation resulted in a proline being replaced by an alanine that leaves a two-carbon gap between the alpha-beta dimer bonds. This gap relaxes the T structure and allows the iron ion to be more in the plane of the porphyrin ring that raises the oxygen affinity of the deoxyhemoglobin and, therefore, allows it to bind oxygen more readily under lower pressures.[1] In other words, the very structure of its blood allows it to carry oxygen more efficiently.

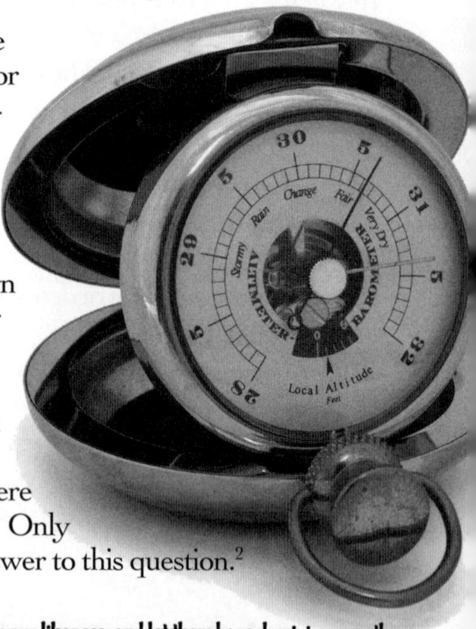


It was found that other high altitude birds have similar changes. Thus, for evolution to be the correct explaina-tion, these changes in hemoglobin must have occurred multiple times. This is an incredible coincidence unless the ability for blood to modify itself was built into the design of these birds. In other words, their DNA code has a built-in ability to adapt in order to fill different envi-ronmental niches, i.e. "programmed filling." The question which evolution has not answered is: "Where did this programming come from?" Only the Bible provides a satisfactory answer to this question.[2]

**And God said, Let us make man in our image, after our likeness: and let them have dominion over the fish of the sea, and over the fowl of the air**

**- Genesis 1:26**

If everyone descended from Adam, where did all the races come from? It began when the people were separated into different language groups at the Tower of Babel. Before this separation, the population shared the same genetic traits; there was a common gene



pool. After the flood, as people scattered out and filled the Earth, specific genes for specific appearances became isolated in separate groupings of people. Races are not the result of new information but the sorting of pre-existing information.

Let's take skin color as an example. Relatively small groups of migrating people would not have carried the same broad range of skin color as the original larger group. One group may have had "darker" genes, while the other "lighter" genes. These small groups would marry within their own language group. As these small groups moved away from Babel, they encountered different climate zones. The people who migrated to the cold regions would experience little sunlight; the dark-skinned members of the group would not have been able to produce as much vitamin D and, thus, would be less healthy and have fewer surviving children. (Vitamin D is needed daily for a healthy body). So, in time, the light-skinned people would dominate these areas of the world.

Conversely, dark-skinned people groups who migrated out of Babel into Africa found more protection from the intense sun rays, while those with lighter skin would easily be affected by more skin problems.

When we isolate people groups, certain characteristics become highlighted. Where did the "races" come from? The tower of Babel, with its instantaneous language barriers, started the process, that led to the rapid development of all the races upon the Earth. In reality, there is no such thing as different "races" of people. We are all descendants of Adam and Eve.

**And He has made from one blood every nation of men...**

**~Acts 17:26**

Have you considered how barbed wired came to be? As the settlers moved out West, there was the need to keep their animals from wandering too far. Throughout time people have used stone or wood to contain animals, but the plains were in short supply of both.

East Texans attempted to solve the cattle containment problem by planting a thorny bush called Osage Orange to contain the herds. These thorny bushes worked well as fencing; however, they were a nuisance to grow and maintain. Michael Kelly noticed the thorns on the bush and thought to copy these thorns onto a wire fencing design. The inventor of barbed wire wrote on his patent, "My invention (gives) to fences of wire a character approximating to that of a thorny hedge." He called his business the Thorn Wire Hedge Company.

Barbed wire changed the life on the Great Plains dramatically – allowing greater productivity. This one invention allowed for the rapid expansion of the West, feeding of millions within our cities, and economic expansion of our country. All because the inventor of barbed wire recognized a brilliant design from nature and copied it for our benefit. We only need to open our eyes to see how the great Designer has laid before us solutions to the dilemmas of life.

**It is the glory of God to conceal a thing: but the honour of kings is to search out a matter.**

**~Proverbs 25:2**

In 2003 the human genome project was completed and the three billion letters on the human genome (DNA code) were mapped in a massive, coordinated effort by genetic research labs around the world. Our biological understanding has exploded in these ten years such that there is no reason to attribute life to slow evolutionary processes. The explanatory power of Darwinism has plummeted as knowledge concerning the complexity of life's design has skyrocketed.

No knowledgeable genetic researcher continues to refer to "junk" DNA. The faulty assumptions of evolution slowed the discovery of DNA functions for years because if we evolved from previous forms of life, there should be vast stretches of the coded molecule that have no functional purpose. So why look for a purpose? Yet, exactly the opposite has happened. Every part of the DNA code, whether directly involved in protein coding or not, is being revealed to have some useful function. This is the opposite of what evolution would predict.

God, who desires that no one be deceived, is making it absolutely crystal clear that recent creation, exactly as described in the Bible, is a reality.

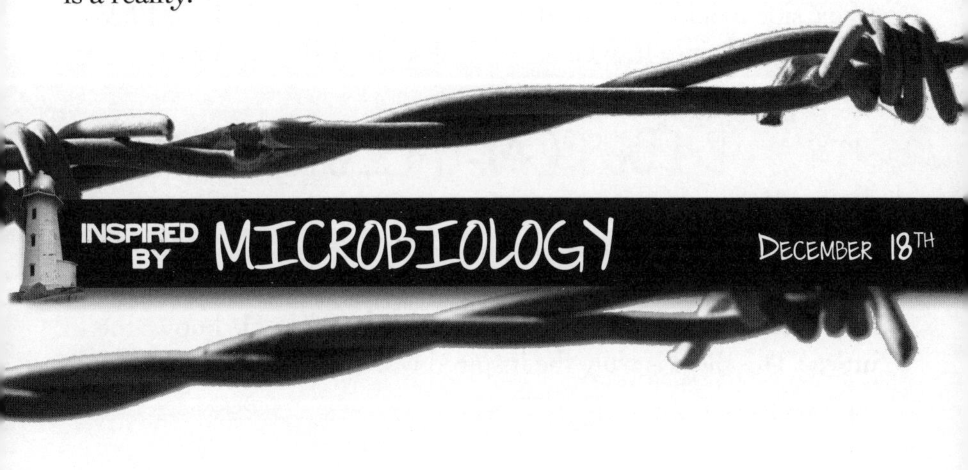


...the thunder of his power who can understand?

- Job 26:14b

The Old Testament has hundreds of specific prophecies concerning Jesus Christ. These prophecies were spoken of hundreds of years prior to his birth. Here are but a few:

| | OT | NT Fulfilled |
|---|---|---|
| Virgin birth | Isaiah 7:14 | Matt. 1:23 |
| Birth in Bethlehem | Micah 5:2 | Matt. 2:6 |
| Fleeing to Egypt | Hosea 11:1 | Matt. 2:15 |
| Crying in Ramah | Jeremiah 31:15 | Matt. 2:18 |
| John's voice in the wilderness | Isaiah 40:3 | Matt. 3:3 |
| Holy Spirit descending on Jesus | Isaiah 11:2, 61:1 | Matt. 3:16 |
| Blind receive sight | Isaiah 35:5 | Matt 11:5 |
| Jesus' 3 days & 3 nights buried | Jonah 1:17 | Matt. 12:40 |
| Speaking in parables | Psalm 78:2 | Matt: 13:35 |
| Jesus' entry on donkey | Zechariah 9:9 | Matt. 21:5 |
| Jesus would cleanse the Temple | Malachi 3:1 | Matt. 21:12 |
| Jesus betrayed for 30 silver pcs. | Zechariah 11:13 | Matt. 27:9-10 |
| Dividing Jesus' garments by lots | Psalm 22:18 | Matt. 27:35 |
| Crucified between two robbers | Isaiah 53:12 | Matt. 27:38 |
| Jesus buried in rich man's tomb | Isaiah 53:9 | Matt. 27:57-60 |
| Not a bone would be broken | Ps. 34:20 | John 19:33 |
| Jesus' side would be pierced | Psalm 22:18 | John 19:23 |
| Jesus ascends into heaven | Psalm 68:18 | Acts 1:9 |



No ordinary book would have hundreds of prophecies fulfilled. This is God's method of validating that the Bible was inspired for our benefit by the One who made time. Therefore He knows the future. The Bible is truly the inspired word of God!

**[Jesus Christ] now is made manifest, and by the scriptures of the prophets, according to the commandment of the everlasting God, made known to all nations... ~ Romans 16:26**

Evolutionists have a foundational principle through which they view geological evidence—the uniformitarian principle. Their key belief can be expressed, "The present is the key to the past". Evolutionists observe what they see today and extrapolate into the past. They believe that slow, gradual processes have created the geology we see today. What we actually observe, even today, are geological catastrophes such as tsunamis, volcanoes, and mud slides altering the landscape. The most devastating and unique catastrophe that reshaped the world was the Flood of Noah's day. This one event created:

- The sedimentary rock that covers three-fourths of the Earth's surface.
- The fossil record: a worldwide graveyard containing trillions of dead plants and animals laid down in layers all over the world.

## INSPIRED BY GEOLOGY — DECEMBER 20TH

- Fossils on every major mountain range in the world including the Himalayas.
- Mountains and our current continents as a result of rapid and widespread land movements during the flood event.
- An entirely new sea floor.
- A subsequent worldwide Ice Age.

What we really should be saying is "The past is the key to the present" not "The present is the key to the past". Isn't it just like Satan, to twist the truth backwards? The year-long flood of Noah completely resurfaced planet Earth. This flood, written about in Genesis, explains the geology we see today. "The past is the key to the present!" The Bible, not the uniformitarian theory, is correct.

***..... There shall come in the last days scoffers, walking after their own lusts, and saying, "Where is the promise of his coming? For since the fathers fell asleep, all things continue as they were from the beginning of the creation."***

***~ 2 Peter 3:3,4***

There are eight planets (Pluto has been downgraded to a planetoid) with more than 63 moons in our solar system, yet Earth is the only place to observe a total solar eclipse. Is this by chance or purpose? A solar eclipse takes place when the moon gets between the path of the sun and the Earth, causing the sun to be blocked out. The moon's size and distance is wondrously perfect to make this event possible! The sun is almost exactly 400 times larger than the moon and also exactly 400 times further away. Because of this, the moon fits perfectly over the sun, allowing us to see, study, and understand the sun's atmosphere. That's an incredible "coincidence."



This "coincidence" has resulted in great scientific discoveries that would have otherwise been impossible. Here are three:

- Perfect solar eclipses allow astronomers to learn about the nature of stars. Using spectroscopes, they have examined how the sun's color spectrum is produced – allowing them to interpret the composition of distant stars.
- Perfect solar eclipses confirmed the fact that gravity bends light.
- Perfect solar eclipses provided a history record that enabled astronomers to calculate the change in the Earth's rotation over the past several thousand years. Because of this, ancient cultural events that recorded the time of solar eclipses can be pinpointed by our modern calendar system.

If the moon had been larger or smaller, or closer or farther, none of this would have been possible. This is not a coincidence. God wants us to explore and learn from the heavens He has made!

*When I consider thy heavens, the work of thy fingers,*
*the moon and the stars, which thou hast ordained..*
*~ Psalm 8:3*

The conch is one tough shell. Yet, it is made out of aragonite, a very brittle mineral. The shell is tough because of its design structure. It has three layers of rod-like structures that crisscross at 90 degrees. The rods are held together with protein glue, making the shell 99% mineral and 1% protein. Each crisscrossing rod is made up of smaller rods and these rods are made up of still smaller rods and so forth, on down to the individual crystals. This arrangement makes it difficult for a crack to travel through the structure. This design arrangement makes the brittle mineral 30 times stronger and about 1,000 times tougher than the pure mineral.



If this was not amazing enough, a living conch can repair itself. Within 24 hours of a wound, a transparent membrane is made. Then, tiny aragonite crystals are deposited, forming many fine layers. After 6-8 days, the conch deposits the elongated crystals followed by the amazingly tough cross layering. The repair process requires fine coordination of mineral and organic layers.

When we examine the repair mechanism programmed within the shell; when we examine the detail design of the shell; when we examine the living creature within the shell… we know there must be a Programmer, Designer, and Originator who created this remarkable creature.

*Praise him for his mighty acts: praise him according to his excellent greatness.*
*~ Psalm 150:2*

The Earth is a very, very, very rare planet because it is capable of supporting life. There are many factors needed to have a habitable planet; here are a few:

- **Liquid water**: The chemical properties of water are perfect for carbon-based life. It allows chemicals to be dissolved and transported. Water also absorbs heat from the sun and regulates temperature.
- **Liquid iron at its core**: The liquid iron generates our protective magnetic field which holds our atmosphere in place.
- **Nitrogen, oxygen, carbon dioxide atmosphere:** The atmosphere is less than one percent of the diameter of the planet, yet it gives Earth a temperate climate, protection from



the sun's radiation and the correct combination of gases for life.
- **Moon:** We have a large moon compared to the size of our planet, stabilizing our axis tilt and driving our tides; both are critical for life.
- **Sun:** Our sun is a main sequence G2 dwarf star. If our sun were less massive, like 90% of the stars in our galaxy, life would not be possible.

All together, there are at least 20 factors making our planet optimized for habitability. The probability of all these factors being present at the same time and place are astonishingly small. The probability of these factors coming together on one planet is less than one in $10^{15}$. Since the vast majority of stars do not even have planets, it is essentially impossible that any other planet in the universe has life. Our Earth is rare indeed.

**I am the LORD that maketh all things; that stretcheth forth the heavens alone; that spreadeth abroad the earth by myself ~ Isaiah 44:24**

During the Middle Ages when many people did not know how to read, the church held plays to teach biblical truths. The play

## INSPIRED BY HISTORY

DECEMBER 24TH

performed during the Christmas season was always about Adam and Eve. What do Adam and Eve have to do with Christmas?

The play started with apples tied to an evergreen tree. There were no apple trees in cold, wintry Germany, so they used an evergreen tree with apples. Adam and Eve chose rebellion against God and ate one of the apples. Immediately, Adam and Eve knew their relationship with God had changed and attempted to hide themselves, but God came looking for them. God had to remove them from the beautiful garden. Sin has been with us ever since. That is why we have death in the world; for the wages of sin is death (Romans 6:23). But how can sin be paid for? Who could make us clean again? Only one – God's son.

Jesus' birth, which is celebrated at Christmas time, is all about paying the debt of Adam and Eve (and ours). Jesus came and paid that price with His death on the cross. Why a tree at Christmas?... to represent that first tree of life. Why an apple on the tree?...to remind us of our sinful choices. Why lights on a Christmas tree?... to signify the coming of the light of the world. Why are Christmas colors green, red, and white?...because of the tree of life (green), the forbidden fruit (red), and the light of the world (white). Christmas traditions, when properly understood, point to the profound truths of Christianity.

These Christians of years-gone-by can help Christians today understand the connection between the Old Testament and the New Testament.

**I am come a light into the world, that whosoever believeth on me should not abide in darkness.**

**~ John 12:46**

The scientific name for mistletoe is Phoradendron, meaning "tree thief". Mistletoe is commonly used as a Christmas decoration, popularized by its bright white berries and green foliage during the Christmas season when other vegetation has turned brown. The custom of kissing under mistletoe may be a carry-over from an older custom requiring enemies who met under a clump of mistletoe (whose extract is extremely poisonous) to disarm, embrace, and refrain from combat. Mistletoe is actually an agent of death in more ways than having poisonous leaves. It is a parasite which burrows its wedge-shaped roots into the limbs of host trees and gradually sucks all of the moisture and nutrients out of its host. With each passing year, the tree weakens and produces fewer leaves while the mistletoe spreads and produces more leaves. Eventually the only green left upon the tree are the mistletoe leaves. Once this happens, the mistletoe also dies. Why would evolution develop a process which automatically results in the organism's own death?

Parasites exist only because we live in a fallen, death-filled, sin-influenced world. Plants such as mistletoe would not be expected from evolutionary processes, but they testify to the reality of an intelligent designer who allowed death to enter creation because of mankind's actions. Yet, this same Designer entered into creation as a babe in a manger, Jesus Christ, in order to conquer death by taking that penalty upon Himself. He then rose victorious over death. This is the real message and meaning of the Christmas season.

**Now when Jesus was born in Bethlehem of Judea in the days of Herod the king, behold, there came wise men from the east to Jerusalem...And they said unto him, in Bethlehem of Judea: for thus it is written by the prophet.**

**~Matthew 2:1, 5**

## Inspired by Biology

December 26th

Have you considered the migration of the black poll warbler? This crazy bird flies in the wrong direction when migrating! It starts its long journey by flying eastward, out over the Atlantic Ocean, before heading south to South America. Why not fly directly to South America?

It somehow knows to fly toward Africa at the exact time needed to catch higher winds that aid in its journey south. The warbler then relaxes and flies to Brazil with a minimum amount of effort. Young warblers kept in captivity with no training from their parents follow the same unusual path. How do they know how to do this?

Many other small songbirds from northeastern USA also fly over the Atlantic Ocean on their way to South America. Why don't they take the land route? A narrow strip of Central America is the primary migration route for millions of raptors. These predator raptors need to fly on thermal air currents. The open ocean does not produce these needed thermals. So by flying out over the Atlantic Ocean, songbirds are safe from these predators. How did they know this? How do these birds know when to leave, what route to take, and where to go? It is written within their DNA code to follow this route at this particular time. It had to work the first time because a drowned or eaten songbird is of no use to the future of the species.

Black poll warblers aren't so crazy after all!

**Oh that I had wings like a dove! for then would I fly away, and be at rest. ~ Psalm 55:6**

Have you ever thought about how the evergreen survives the winter? For an "evergreen" to stay green forever, the needles must carry on photosynthesis throughout the winter. Water is needed for this to happen. In the winter, the ground is frozen, so where does the water come from?

To keep the water inside for as long as possible, the needles have a special design. Evergreens have narrow needles with a thick waxy coating. This coating helps retain water. The small breathing holes (stomata) under the needles are few and slightly sunken, which again protects the needles from losing too much water. But what about those frigid temperatures – wouldn't the water freeze inside the needles?

The needles themselves produce certain chemicals that act like anti-freeze, preventing ice crystals from forming inside the cells. These are also the chemicals that give the evergreens their characteristic scent. How many thousands of years would it have taken to develop these anti-freeze chemicals by random mutations? How did an evergreen survive until this happened? Just how did the tree evolve these complex "anti-freeze" chemicals? Could this have happened by accident and chance?

The Master Designer had to set it up correctly from the very beginning. Otherwise, all that would have "evolved" would have been dead evergreens. So, as you take that winter walk, marvel at the design features of the narrow, waxy needles and its specially formulated anti-freeze allowing the evergreen to survive the winter.

**I will plant in the wilderness the cedar, the shittah tree, and the myrtle, and the oil tree; I will set in the desert the fir tree, and the pine, and the box tree together: That they may see, and know, and consider, and understand together, that the hand of the Lord hath done this, and the Holy One of Israel hath created it. - Isaiah 41:19-20**

# INSPIRED BY BOTANY

DECEMBER 28TH

Evergreen trees are called "ever-green" because their needles are green year-round. The evergreen's thin needles are also designed to shed ice and snow easily so that the enormous weight of snow does not break the evergreen branches. The branches are able to withstand these snow loads due to their cone-shape or "christmas-tree shape." Each year these spruce and fir trees produce a new whorl of three to six branches that grow horizontally like spokes from the hub of a wheel. Each branch is flexible, which allows it to bend gently down when there is pressure from the snow. As the snowy branch bends down, it is supported by last year's branch beneath it. Finally, as the pitch increases, the snow slides right off.

Deciduous trees do not have this type of branch flexibility or cone shape to provide a supporting framework. It seems as though God made evergreen trees to remind us of life and hope throughout the winter season.

**The glory of Lebanon shall come unto thee, the fir tree, the pine tree, and the box together, to beautify the place of my sanctuary...**

**~ Isaiah 60:13**

All coal mines serve as vivid reminders of Noah's Flood. Inside a coal mine are the remains of leaves, trees, shells and sometimes fish. Going underground is like traveling into a long lost world. Coal miners frequently uncover carbonized branches and roots; in some places, the brush and trees are heaped up as if they were dumped near a flooding river. In other places, tree trunks as long as 80 feet are found, some lying down and others extending upward through the seams of coal. The coal seams or layers are often sandwiched between layers of sandstone, which were laid down by flowing water.

When the machinery working coal seams is turned off, methane can often be heard hissing from the coal face. Methane gas is explosive and poisonous. For the methane to still be present suggests coal is not millions of years old; otherwise, the gas would have leaked out long ago.

Miners have to be continually aware of deadly stump and tree roots called "kettle bottoms", which are petrified in the sandstone seam. As miners are digging out the coal seam, the sandstone roof may contain these heavy, petrified tree roots. These heavy kettle bottoms sometimes fall out without warning and hurt the miners. Miners working deep in the Earth see evidence of a global flood in the coal, sandstone, methane gas, and fossils. The existence of widespread coal seams is a testimony to the catastrophic worldwide flood of the Bible.

***For yet seven days, and I will cause it to rain upon the earth forty days and forty nights; and every living substance that I have made will I destroy from off the face of the earth. ~ Genesis 7:4***

Many people have come to believe that hell is just a fictitious fantasy. Yet, Jesus referred to this as a very real place dozens of times. So just what, or where, is hell? Many theologians define hell as the total and complete absence of God. Even though we



refer to things happening here on Earth in terms of "hell on Earth" or "war is hell", it is readily apparent that God's presence and guiding hand is at work all around us. It is a frightening idea, the concept of the total removal of God's presence. This is because God is love. Thus, hell will be the total removal of love from our existence. So, why would God allow this?

For love to exist, there must be freedom. You cannot force your children or spouse to love you; they must have the freedom to walk away or stay. God gives us the same freedom, yet the marvel of God's love is that He continues to love us even when we turn our back on Him. But once we die, our decision of whether or not to love God has ended. At that point God will grant us our desire and there are only two choices:

(1) Accept what He has done to bring us back into fellowship with Himself by accepting the death of Jesus as payment for our rebellion.

(2) Live our lives for ourselves, our desires, our comfort, and our own good works.

If we choose option two, God will indeed grant us our wish, and we will be forever separated from the presence of perfect love. We ourselves, not God, will have cast ourselves into hell.

**And as it is appointed unto men once to die, but after this the judgment.**

**~Hebrews 9:27**

How did kangaroos get to Australia if they disembarked the Ark in Turkey after Noah's Flood ended? Noah's Flood began when the "fountains of the deep" burst forth. Massive volcanism and rapid continental movement during the flood would have increased the ocean water temperature. Higher ocean temperatures would have resulted in increased evaporation, thus providing the moisture for the massive snowfalls driving the Ice Age. These were the conditions upon the Earth immediately following the Flood.

## December 31st

## Biology Inspired By

Michael Oard, in his book *Frozen in Time,* has calculated it took 500 years for the ice to build up and 200 years for it to melt back. Enormous amounts of water became locked up in glaciers and ice sheets. This would have lowered the ocean levels and animals could have easily migrated across the land bridges to Australia and to North America. As the ice age ended, the sea levels rose hundreds of feet, and the land bridges were flooded, sealing the unique animal life of Australia, such as kangaroos, onto the island. The Ice Age and animals migrating off the Ark to all parts of the world are not a mystery; these things are direct consequences of a real worldwide flood and true biblical history.

**Moreover I will appoint a place for my people Israel, and will plant them, that they may dwell in a place of their own, and move no more...**

**- 2 Samuel 7:10**

# REFERENCES


1-Jan Jason Lyle, *The Ultimate Proof of Creation: Resolving the Origins Issue*, (Master Books, 2009), p.45.
2-Jan Andrew Snelling, *Noah's Flood and the Earth's Age*, 2009.
3-Jan www.snowcrystals.com
5-Jan Donald DeYoung and Derrick Hobbs, *Discovery of Design*, 2009, pp. 112-113, 209
7-Jan John Hudson Tiner, *Exploring the History of Medicine*, 2001, pp.68-74.
8-Jan *Jonathan Park: The Hunt for Beowulf*, volume IV, study guide, p.45-51
9-Jan Donald DeYoung, *Weather and the Bible*, pp. 89-90.
10-Jan 1. Donald DeYoung, *Weather and the Bible*, 1992, pp. 53-55.
2. http://www.kidsastronomy.com/the_planets.htm
11-Jan *Considering God's Creation*, p.39.
12-Jan Felice Gerwitz and Jill Whitlock, *Creation Anatomy*,1996, pp.12-13.
13-Jan John Morris, *The Young Earth*, (Master Books, 2009), pp.54-55.
13-Jan John D. Morris, *The Geology Book*, (Master Books, 2000), p. 33-34.
15-Jan http://www.creationmoments.com/content/cyanide-defense-0
16-Jan 1. Donald DeYoung, *Dinosaurs and Creation*, 2000, pp.46-47.
2. Richard and Tina Kleiss, *A Closer Look at the Evidence*, 2003, p. 61.
19-Jan 1. Mary H. Schweitzer, et al., "Soft-Tissue Vessels and Cellular Pres. in Tyr. rex", *Science*, 307:1952-1955, 2005.
2. Schweitzer, M.H., et al., "Biomolecular characterization and protein sequencing of the Campanian hadrosaur", B. Canadensis, *Science*, 324:626-631, 2009.
20-Jan Ken Ham, "Do I Believe in UFO's? Absolutely!", *Answers Magazine*, Jan.-Mar. 2008, pp. 60-62.
21-Jan 1. Creation moments, "The Never-Fail Alarm", www.creationmoments.org.
2. Joanne De Jonge, *Silent Signals and Secret Codes*, The Banner.
23-Jan www.setterfield.org
24-Jan Jonathan Sarfati, , "Diamonds: a creationist's best friend", *Creation Magazine*, September-November 2006, pp 26-27.
25-Jan Lee Strobel, *The Case for a Creator*; 2004, pp. 185-186.
26-Jan Joanne E De Jonge, *It's God's World Magazine*, "The Inside Story", 1/6/95, pp. 6-7.
27-Jan E.A. Widder, "Bioluminescence in the Ocean", *Science*, 328:704-708, 2010.
28-Jan Kevin May, "Born to Communicate", *Creation Magazine*, March-May 2009, pp. 40-42.
29-Jan AstroBiology http://www.astrobio.net/exclusive/2419/our-earliest-animal-ancestors, 2010.
30-Jan Book Jonathan Sarfati, "The Greatest Hoax on Earth?: Refuting Dawkins on Evolution", (Creation Publishers, 2010), pp. 233-248.
31-Jan 1. http://creation.com/fibre-optics-in-eye-demolish-atheistic-bad-design-argument
2. David Menton, *The Hearing Ear and Seeing Eye*, DVD.
1-Feb John Phillips, *Exploring Genesis*, 1980, p. 44.
2-Feb Stuart Burgess, "Hallmarks of Design", p. 56-59, 2004.
3-Feb 1. Woolly Bear Caterpillars Can Freeze and Thaw, *Discover Mag.*, teacherweb.puyallup.k12.wa.us/.../ woolly_bear_caterpillars_can_freeze_and_ thaw.doc
2. teacherweb.puyallup.k12.wa.us/phs/rstanley/documents/woolly_bear_caterpillars_can_freeze_and_thaw.doc
4-Feb Gordon Wilson, *Answers magazine*, "Fungus Firearms", p.26-28, April-June 2010
5-Feb Donald B. DeYoung, *Weather and the Bible*, 1992, p.58.
6-Feb Donald DeYoung, *Astronomy and the Bible*, 2000, pp. 89-90.
7-Feb 1. Paul Bartz, *Letting God Create Your Day*, "Birds Egg Evolution", p.67.
2. http://en.wikipedia.org/wiki/Common_Murre#Eggs_and_Incubation
8-Feb Donald Chittick, *The Puzzle of Ancient Man*, 2006, pp. 103-105, 127-132.
9-Feb 1. *Creation*, 9-11-99, pp. 54-55, "Walking Trees."
2. http://creation.com/walking-trees
11-Feb Doreen Cubie, "The Better to Hear You With", *National Wildlife*, Dec./Jan. 2010, p.24.
12-Feb AstroBiology Magazine - http://www.astrobio.net/exclusive/2419/our-earliest-animal-ancestors, 2010.
16-Feb 1. Chip and Dan Heath, *Made to Stick*, (Random House, 2007), p.68.
2. Dr. Jerry Bergman, *The Dark Side of Charles Darwin*, (Master Books, 2011), pp. 170-188.
17-Feb http://www.create.ab.ca/superior-farms/
18-Feb Charles Edwards, *God by the Numbers*, http://christianitytoday.com/ct/2006/003/26.44.html NRT
20-Feb Werner Gitt, "Counting the stars: The vastness of the universe is cause for joy, not loneliness", *Creation* 19(2): 10-13, March 1997.
21-Feb http://creation.com/a-birdbox-and-a-tree
22-Feb Daniel G. Gibson, et al, "Creation of a Bacterial Cell Controlled by a Chemically Synthesized Genome", *Science*, 329:52-56, 2010.

23-Feb Alexander Williams, "God's Amazing Glue", *Creation Magazine,* March-May 2002, p.27.
24-Feb 1. Paul Bartz, "This Carbon-14 Discovery is a Gem", *Creation Moments.*
2. http://www.creationmoments.com/content/carbon-14-discovery-gem
25-Feb www.creationontheweb Grass-eating dinos
26-Feb Dogs Decoded: Understanding the Human-Dog Relationship, NOVA broadcast, BBC, 2010.
28-Feb 1. Paul Bartz, "The Warm-Blooded Bumblebee", *Letting God Create your Day,* p. 245.
2. http://www.bumblebee.org/bodyTempReg.htm
3. http://www.bumblebee.org/lifecycle.htm
2-Mar 1. *Jonathan Park: The Explorer's Society study guide,* p. 10-23.
2. John Morris, *The Young Earth,* p. 45-46, 2007.
3-Mar 1. Jerry A. Coyne, *Why Evolution is True,* (Viking 2009), p. 80.
2. http://www.med.umich.edu/lrc/coursepages/M1/embryology/embryo/18changesatbirth.htm
4-Mar Philip Snow, *The Design and Origin of Birds,* 2006, 101-107.
5-Mar D. James Kennedy, *What If Jesus Had Never Been Born?,* 1994.
6-Mar *Biology: God's Living Creation,* Abeka, 1997, p.65-66.
7-Mar John Phillips, *Exploring Genesis: An Expository Commentary,* 1980, pp. 42-43.
8-Mar Dennis Peterson, *Unlocking the Mysteries of Creation,* pp. 46-59, 2002.
11-Mar Ron Carlson and Ed Decker, *Fast Fact on False Teachings,* 1994, pp.64-65.
12-Mar Ron Carlson and Ed Decker, *Fast Fact on False Teachings,* 1994, pp.64-65.
13-Mar *101 Scientific Facts and Foreknowledge,* p.25.
14-Mar Joanne De Jonge, "Silent Signals and Secret Codes", *The Banner,*
15-Mar Wood Petrified in Spring, *Creation Magazine,* June-August 2006, 18-19.
16-Mar Wayne Ranney, *Carving Grand Canyon: Evidence, Theories, and Mystery,* Grand Canyon Association, 2005.
16-Mar Tom Vail, et al., *Your Guide to the Grand Canyon,* (Master Books, 2008), p. 143.
17-Mar D. James Kennedy, "St. Patrick's Purpose-Driven Life", March 2007, p.8.
18-Mar Harun Yahya, *Atlas of Creation vol. 1,* 2006, p.732.
19-Mar Donald B. DeYoung, *Weather and the Bible,* 1992, pp. 77-78.
20-Mar Heather M. Brinson, "The Human Body-Wired for Extremes", *Answers Magazine,* p.36-39, Oct.-Dec. 2009,
21-Mar *Northwoods Companion,* p.75
22-Mar Dennis R. Petersen, *Unlocking the Mysteries of Creation,* 2002, pp. 224-225.
26-Mar Donald DeYoung and Derrik Hobbs, *Discovery Design: Searching Out the Creator's Secrets,* (Master Books, 2009), pp.48-49.
27-Mar Margaret J. Helder, *Completing the Picture,* 1990, p.13-19.
28-Mar http://creation.com/age-of-the-Earth
29-Mar Lanny and Marilyn Johnson, "More than Just a Nose", *Alpha Omega Institute newsletter;* September-October 2006.
30-Mar Paul Bartz, "Air-cooled Elephants", *Letting God Create Your Day,* volume 4, p. 97.
31-Mar Donald DeYoung and Derrik Hobbs, *Discovery Design: Searching Out the Creator's Secrets,* (Master Books, 2009), p. 36.
1-Apr Jonathan Sarfati, *The Greatest Hoax on Earth?: Refuting Dawkins on Evolution,* (Creation Book Publishers, 2010), p.43.
3-Apr http://en.wikipedia.org/wiki/Bamboo
4-Apr Dr. Jay Wile, Exploring Creation with General Science, pp. 286-289.
5-Apr 1. David Catchpoole, "Why a butterfly flutters", Creation Magazine, March 2004, p. 56.
2. http://creation.com/why-a-butterfly-flutters-by.
6-Apr David Catchpoole, "Death throes", Creation 31(3) June-August 2009, pp. 42-43
8-Apr Personal correspondence with Dr. Willie Dye, president, New Covenant Institute of Biblical Archaeology.
9-Apr 1. George Mulfinger & Julia Mulfinger Orozco, *Christian Men of Science: Eleven Men Who Changed the World,* 2001, p. 202.
2. Spike Psarris, *Our Created Solar System,* http://www.creationastronomy.com
10-Apr 1. Jobe Martin, *Incredible Creatures That Defy Evolution II,* DVD.
2. Philip Snow, *The Design and Origin of Birds,* 2006, pp.17-18.
11-Apr Andrew Snelling, "Radiohalos: Startling evidence for catastrophic geological processes on a young Earth", *Creation Magazine,* March-May 2006, pp.46-50.
12-Apr Andy McIntosh, "100 years of Airplanes-but these weren't the first flying machines!", Creation Magazine, December 2003, *Creation,* 26 (1):44-48.
13-Apr Jeannie K. Fulbright, *Exploring Creation with Zoology 2: Swimming Creatures of the Fifth Day,* 2006, pp.133-134.

14-Apr James Johnson, "Fighting over the Furniture and Faith", *Acts & Facts*, Vol.39 No.12, Dec. 2010, p. 8,9.

15-Apr http://www2.mcdaniel.edu/Biology/eco/fire99/Fireweb.htm

16-Apr Dr. Don Bierle, *Surprised By Faith*, 1992, pp. 37-39.

17-Apr Dale Tacket, *The Truth Project* ,Session #10, Focus on the Family, 2008.

19-Apr Philip Snow, *The Design and Origin of Birds*, p.17.

21-Apr http://creation.com/focus-creation-news-and-views-291

22-Apr Donald DeYoung and Derrik Hobbs, *Discovery Design: Searching Out the Creator's Secrets,* (Master Books), 2009, p.120.

23-Apr 1. Wayne Jackson, *The Human Body: Accident or Design,* p. 73-75, 2000.
2. *Biology: God's Living Creation*, A Beka Book, p. 288-291, 1997.

24-Apr Jonathan Sarfati, "Mars: the Red Planet", *Creation Magazine*, 32(2), p. 38-41, 2010.

25-Apr http://www.answersingenesis.org/articles/am/v4/n4/heart

26-Apr *Dogs Decoded: Understanding the Human-Dog Relationship,* NOVA broadcast, BBC, 2010. NRT

27-Apr Bruce Malone, *Censored Science: The Suppressed Evidence,* (Search for the Truth Publications, 2009), pp.72-73.

28-Apr Ruth Beechick, *Genesis Finding our Roots*, 1997, pp. 13, 16.

29-Apr Myles Willard, *The Rest is History,* monograph, Mayville, MI, 2008.

30-Apr Lanny Johnson, "Fibonacci Numbers and the Divine Proportion", *The Hand of God*, 2009.

1-May Donald DeYoung and Derrik Hobbs, *Discovery Design: Searching Out the Creator's Secrets,* (Master Books, 2009), p. 14.

2-May John Phillips, *Exploring Genesis: An Expository Commentary,* 1980, pp. 43-44.

3-May Donald DeYoung and Derrick Hobbs, *Discovery of Design*, 2009, p. 50-51.

4-May Paul A. Bartz, *Letting God Create Your Day*, vol.4, p.221.

5-May *Biology: God's Living Creation,* A Beka, 1997, p.21.

6-May 1. http://www.creationmoments.com/content/ancient-meteorologists-predicted-el-ni%C3%B1o-0
2. Paul Bartz , "Ancient Meteorologists Predicted El Niño", *Creation Moments radio program*

7-May 1. Michael and Beverly Oard, *Life in the Great Ice Age*, (Master Books, 1993), p. 60.
2. James I. Nienhuis, *Ice Age Civilizations*, 2006, pp. 3-6.

9-May Michael Oard, *Frozen in Time: The Woolly Mammoth, The Ice Age, and the Bible,* 2004, p.33, 40-44.

10-May Stuart Burgess, *The Origin of Man: The image of an ape or the image of God?*, 2004, pp.64-70.

11-May 1. John Phillips, *Exploring Genesis*, 1980, p.41.
2. Mark Breen and Kathleen Friestad, *The Kid's Book of Weather Forecasting,* 2000, p.79.

12-May Del Tackett, *The Truth Project,* DVD #5, Focus on the Family, 2007.

13-May Lee Strobel, *The Case for Creation*, p. 175-178, 2004.

14-May John C. Sanford, *Genetic Entropy and the Mystery of the Genome*, (FMS Publications, Third Ed., 2008).

15-May John Woodmorappe, "The Dracula Connection to a Young Earth", *Creation Magazine,* December 1998, p. 32. Creation 21(1):32

16-May 1. Kyle Butt and Eric Lyons, *Dinosaurs Unleashed*, 2004, pp.2-3.
2. Ken Ham, *The Great Dinosaur Mystery*, 1998, pp.94-96.

17-May Ken Ham, *The New Answers Book*, 2006, p.149-176.
Ken Ham, *The Great Dinosaur Mystery*, 1996, p.39.

18-May http://www.creationmoments.com/content/worlds-strangest-bird-0

19-May 1. Henry M. Morris, *The Genesis Record*, 1976, pp.22-30.
2. Ruth Beechick, *Genesis Finding our Roots*, 1997, p.28.

21-May 1. Monty White, "The Amazing Stone Bears of Yorkshire", *Creation Magazine*, June-August 2002, pp. 48-49.
2. Carl Wieland, "The Earth: How Old Does it Look?", *Creation Magazine,* Dec. 2000, pp. 8-13.

22-May http://www.blurtit.com/q749905.html

24-May http://www.learner.org/jnorth/tm/loon/AdaptationsHead.html

25-May 1. Andrew Snelling, *Earth's Catastrophic Past*, 2009, pp. 529-530, 1082.
2. Andrew Snelling, *Grand Canyon: Testimony to the Biblical Account of Earth's History*, 2009, DVD

26-May 1. Andrew Snelling, "No Slow and Gradual Erosion", *Creation Magazine*, Jan-March 2009, pp. 96-99.
2. John D. Morris, *The Young Earth,* 1994, pp. 96-98.

27-May *Privileged Planet*, DVD

28-May Joanne E. De Jonge, *Rustling Grass,* p.79-83, 1985.

29-May Heather M. Brinson, "Heart Constantly Beating Death", *Answers Magazine*, 9/09.

30-May Steven Austin, *Grand Canyon Monument to Catastrophe*, 1994, p.158-159."

31-May 1. Ruth Beechick, *Genesis Finding our Roots*, 1997, p.55."

31-May 2. Bruce Malone, *Search for the Truth: Changing the World with the Evidence for Creation*, 2006, p.65.

1-Jun 1. www.nwf.com
2. www.nationalgeographic.com

2-Jun "Post -Flood World", *Answers Magazine*, Oct.-Dec. 2008, pp. 66-67.

3-Jun 1. Don De Young , *One Leg Up on Architects,* September 2, 2009.
2. http://www.answersingenesis.org/articles/am/v4/n4/architects

4-Jun John Morris, *The Young Earth*, 2007, pp. 74-75

6-Jun http://www.answersingenesis.org/docs/4005.asp

7-Jun www.nwcreation.net/noahlegends

8-Jun Randy J. Guliuzza, P.E., M.D. ICR newsletter, *Acts and Facts*, 10/10-1/11 and 5/11.

10-Jun teacherweb.puyallup.k12.wa.us/.../woolly_bear_caterpillars_can_freeze_and_thaw.doc

11-Jun Melinda Christian, "Purring Cats and Roaring Tigers", *Creation Magazine*, Oct.-Dec. 2007, pp. 20-22.

12-Jun Paul A. Bartz, *Letting God Create Your Day*, "Natural Insect Repellent", vol. 1, p.51.

13-Jun George Mulfinger & Julia Mulfinger Orozco, *Christian Men of Science: Eleven Men Who Changed the World*, 2001, pp.71-97.

14-Jun www.setterfield.org

15-Jun J. Budziszewski, *What We Can't Not Know*, Spence Publishing, 2003, pp.158.

16-Jun Paul A. Bartz, "Natural Insect Repellent", *Letting God Create Your Day*, vol. 1, p.51.
Jeannie Fulbright, *Exploring Creation with Astronomy*, 2004, p. 52-55.

17-Jun Paul A. Bartz, "One Generation After the Flood", *Letting God Create Your Day*, Vol. 6, 2010, p.133.

18-Jun Don DeYoung, *Our Created Moon*, DVD.

19-Jun 1. Jeannie Fulbright, *Exploring Creation with Botany*, p.139-140, 2005.
2. Abeka, Biology, .p 91-92.

20-Jun Eric Ludy, *The Bravehearted Gospel: The Truth is Worth Fighting For*, (Harvest House Publishers, 2008), pp.121-123.

21-Jun Andrew A. Snelling, *Answers Magazine*, "The World's a Graveyard", p. 76-79, April-June 2008.

22-Jun http://www.thebanner.org/magazine/article.cfm?article_id=580

23-Jun John Phillips, *Exploring Genesis: An Expository Commentary*, 1980, pp. 49-50.

24-Jun Don DeYoung, "Eyelids-Intermittent Wipers", *Answers Magazine*, Jan-Mar. 2011, pp. 30-31.

25-Jun George Mulfinger & Julia Mulfinger Orozco, *Christian Men of Science: Eleven Men Who Changed the World*, 2001, pp.99-117.

27-Jun Stuart Burgess, *The Origin of Man: The image of an ape or the image of God?*, 2004, pp.38-47.

28-Jun Donald B. DeYoung, *Weather and the Bible*, 1992, p. 40-42.

30-Jun John Woodmorappe, "Caring for the Animals on the Ark", *Answers Magazine*, pp.36-38, April-June 2007.

1-Jul 1. Michael Oard, *Frozen in Time: The Woolly Mammoth, the Ice Age and the Bible*, 2004.
2. James I Nienhuis, *Ice Age Civilizations*, 2006.

2-Jul 1. Bruce Malone, *Censored Science: The Suppressed Evidence*, (Search for the Truth Publications, 2009), pp.76-77.
2. N.P.Edwards, et. al., Infrared mapping resolves soft tissue preservation in 50 million year-old reptile skin, *Proc. R. Soc. B*, published online, 23 March, 2011, rspb.royalsocietypublishing.org.

3-Jul Joe Francis, *Answers Magazine*, "Good Designs Gone Bad", July-Sept. 2009, pp. 32-35.

4-Jul http://learners.gsfc.nasa.gov/mediaviewer/sat_super/

8-Jul *Answers Magazine*, "The Beauty and Wonder of our Universe", Jan.-Mar. 2008, Back page chart.

9-Jul Philip Snow, *The Design and Origin of Birds*, p.165.

10-Jul Michael Oard and Hank Giesecke, "Polystrate Fossils Require Rapid Deposition", *Creation Research Society Quarterly Journal*, March 2007, pp. 232-240.

11-Jul "Beware the Bubble's Burst", Creation Magazine, March -May 2009, pp. 50-51.

12-Jul David A. Menton, "Melanin-Umbrellas of our Skin", *Answers Magazine*, Oct.-Dec. 2010, pp.68-70.

13-Jul Don DeYoung, "A Hole in the Design Argument", *Answers Magazine*, April-June 2010, pp.36-37

14-Jul Tom Vail, et al., *Your Guide to the Grand Canyon*, (Master Books, 2008), p. 154

15-Jul Heather M. Brinson, "Heart Constantly Beating Death", *Answers Magazine*, Oct.-Dec., p.53, 2009.

16-Jul 101 Scientific Facts and Foreknowledge, p. 2

17-Jul Paul A. Bartz, "Vegetarian Piranhas?", *Letting God Create Your Day*, 2010. p. 39.

18-Jul 1. Heather M. Brinson, "The Human Body-Wired for Extremes", *Answers Magazine*, pp. 37-38., Oct.-Dec. 2009.
2. www.answersingenesis.org/articles/am/v4/n4/wired-for-extremes

20-Jul John C. Sanford, *Genetic Entropy and the Mystery of the Genome*, (FMS Publications, Third Ed., 2008).

21-Jul Donald B. DeYoung, *Weather and the Bible*, p.133-134, 1992.

22-Jul Gary and Mary Parker, *The Fossil Book*, 2005, pp. 7-8.

23-Jul Dr. Jerry Bergman, *The Dark Side of Charles Darwin*, (Master Books, 2011), p. 188.

25-Jul Jules H. Poirier, *From Darkness to Light to Flight*, 1995.

26-Jul Jules H. Poirier, *From Darkness to Light to Flight*, 1995.

27-Jul Lee Strobel, The Case for a Creator, 2004, pp. 185-186.

28-Jul Donald DeYoung and Derrik Hobbs, *Discovery Design: Searching Out the Creator's Secrets*, (Master Books, 2009), pp.124.

29-Jul Steven DeVowe "The Amazing Motorized Germ", *Creation*, 27(1):24-25, December 2004.

30-Jul The Planet Earth, "Ice Worlds", *BBC*, 2006.

30-Jul Carl Kerby, What is the Best Evidence that God Created?, DVD, 2003.

31-Jul Paul F. Taylor, "Dinosaur Killer", *Answers Magazine*, pp. 64-66, Jan.-Mar. 2008.

1-Aug Jason Lisle, "Global Warming's Solar Connection", *Answers Magazine*, Vol. 5 No. 3, July 2010, pp. 52-55.
2-Aug Dr. Carl Werner, *Evolution: the Grand Experiment vol. 2 - Living Fossils*, 2008, pp.183-230, quote: 210.
3-Aug Dr. Jonathan Sarfati , *By Design*, p.116, 2008.
4-Aug David Catchpoole, "Bunchberry bang!", *Creation*, 31(2):32-34, March 2009
5-Aug 1. Michael Oard, "Do ice cores show many tens of thousands of years?", *Frozen in Time*, 2004, pp.119-126.
2. http://www.icr.org/article/ice-cores-age-Earth/
3. Jonathan Park: *The Winds of Change*, study guide, pp.43-44.
6-Aug 1. Ruth Beechick, *Genesis Finding our Roots*, 1997, pp. 97-98.
2. http://groups.yahoo.com/group/SEQUOIAMINISTRY/message/5254
7-Aug 1. "The Mystery of Mouse Pyramids", *Creation Moments radio*.
8-Aug 1. "Canyon carved in just three days in Texas flood: Insight into ancient flood events on Earth and Mars", *ScienceDaily*, 7/10/10, http://www.sciencedaily.com/releases/2010/06/100620155748.htm.
2. Michael P. Lamb, Mark A. Fonstad. Rapid formation of a modern bedrock canyon by a single flood event, *Nature Geoscience*, 2010; DOI: 10.1038/ngeo894
9-Aug Alan L. Gillen, *Body by Design*, p.121, 2003.
10-Aug Ray Comfort, *The Atheist Test tract*
11-Aug Dan Breeding, "American Alligator- Sovereign of the Swamp", *Answers Magazine*, p. 22-25, June 2010.
12-Aug 1. Gary Forsythe, *Creation News Update*, Vol. 7. #4, fall 2010, www.cryingrocks.org
2. Bruce Malone, *Censored Science: The Suppressed Evidence*, (Search for the Truth Publications, 2009), pp.76-77.
13-Aug David Butvill, "Teamwork: When Animal Enemies Unite", *National Wildlife*, April/May 2008, pp. 18-20.
14-Aug Larry Vardiman,"A Proposed Mesoscale Simulation of Precipitation in Yosemite Nat. Park with a Warm Ocean", *Proceed. of the 6th International Conf. on Creationism*, Pittsburg, PA, 2008, pp.307-319.
15-Aug John Morris, *Is the Big Bang Biblical?*, p.12-13, 2003.
17-Aug 1. http://en.wikipedia.org/wiki/Genetic_disorder
2. Dr. John C. Sanford, *Genetic Entropy and the Mystery of the Genome*, (FMS Publications, Third Ed., 2006).
19-Aug Dennis Dreves, "Beavers Aquatic Architects", *Creation Ex Nihilo*, vol. 15 no. 2, pp.38-41.
20-Aug 1. Dr. John Morris, "Evaporites and the Flood", *Acts and Facts*, June 2010, pp.17.
2. Andrew Snelling, *Earth's Catastrophic Past*, 2009, pp. 937-938.
21-Aug Bruce Malone, *Censored Science: The Suppressed Evidence*, (Search for the Truth Publications, 2009), pp.60-61.
22-Aug Hua Jin, et. Al., "The Conserved Bardet-Biedl Syndrome Proteins Assemble a Coat that Traffics Membrane Proteins to Cilia", *Cell*, 7:1208-1219, 2010.
23-Aug Jules H. Poirier, *From Darkness to Light to Flight*, 1995, pp. 1-2.
24-Aug Lawrence Richards, *It Couldn't Just Happen*, 1989, p.108-109.
25-Aug Bruce Malone, *Censored Science: The Suppressed Evidence*, (Search for the Truth Publications, 2009), pp.74-75.
26-Aug Jonathan Henry, *The Astronomy Book*, p.37, 1999.
27-Aug Ken Ham, *The New Answers Book 2*, 2008, p. 48.
28-Aug Dr. Jobe Martin, *The Evolution of a Creationist*, 1994, p.123-125.
29-Aug Stuart Burgess, *The origin of man: The image of an ape or the image of God?*, 2004, p.31-33.
30-Aug 1. "Dragonfly Design Tips", *Creation Magazine*, 32(2) 2010, p.51.
2. http://www.mndragonfly.org/defined.html
31-Aug 1. Donald DeYoung and Derrick Hobbs, *Discovery of Design*, 2009, pp. 44-45.
3-Sep 1. David Catchpoole, "Toucan's Beak Beats the Heat", *Creation Magazine*, 32 (4) 2010.
2. http://creation.com/toucan-beak
4-Sep Stuart Burgess, *Hallmarks of Design*, pp. 98-118.
5-Sep 1. http://www.brighthub.com/science/genetics/articles/45268.aspx#ixzz0sGKGVZDd
2. http://www.brighthub.com/science/genetics/articles/45268.aspx#ixzz0sGKMD51s
6-Sep Rick Deighton, *Is the Bible Without Any Errors?*, (Search for the Truth Publications, 2011), p.3.
8-Sep A.M. Labin and E.N. Ribak, "Retinal Glial Cells Enhance Human Visual Acuity", *Phys. Rev. Lett.* 104:158102, 2010.
9-Sep Evidence for a Global Flood Grand Canyon, *Answers Magazine*, fold-out April-June 2011.
10-Sep Jonathan Sarfati and David Catchpoole, " 'Backwards' Comet Perplexes Scientists", *Creation Magazine*, p. 38-39, September-November 2009.
11-Sep Richard Peachey, *Creation Magazine*, "Bypassing the Cracks", June-August 2006, pp. 20-23.
12-Sep Ken Ham, *Demolishing supposed Bible Contradictions*, p. 56, 2010.
13-Sep 1. Jonathan Park: *The Winds of Change, study guide*, p. 62-63.
2. Duane T. Gish, *Dinosaurs by Design*, 1992, p.75
16-Sep 1. Donald DeYoung, *Weather and the Bible*, 1992, pp.85-86.
2. Jonathan Sarfati, "Flighty Flippers", *Creation Magazine*, March -May 2005, p. 56.
17-Sep Donald DeYoung and Derrick Hobbs, *Discovery of Design*, pp. 104-105, 208, 2009."

18-Sep Donald B. DeYoung, *Weather and the Bible,* 1992, pp. 61-63
20-Sep 1. Stuart Burgess, *Hallmarks of Design,* 2004, p.75-84.
20-Sep 2. Stuart Burgess, The beauty of the peacock tail and the problems with the theory of sexual selection, *TJ (now Journal of Creation),* 15(2):94-102, August 2001.
21-Sep Carl Werner, Evolution: the Grand Experiment vol. 2 - Living Fossils, 2008, pp. 57-70, 161-168.
22-Sep Andrew Snelling, *The Flood: The Big Picture of its Mechanism and Resulting Evidences,* DVD, 2009.
24-Sep Ruth Beechick, *Genesis Finding our Roots,* 1997, p. 13.
25-Sep Lee Strobel, *The Case for a Creator,* p. 209-211, 2004.
26-Sep 1. James Perloff, *Tornado in a Junkyard,* p.127.
2. Abeka Biology, p.476, 1997.
27-Sep Carl Werner, *Evolution: the Grand Experiment vol. 2* - Living Fossils, 2008. p.1-56.
28-Sep 1. David Noebel & Chuck Edwards, *Thinking Like a Christian,* 1999, p.44.
2. Don Bierle, *Surprised By Faith,* 1992, pp.28-34.
29-Sep *A Beka Science,* grade 4, 1989, pp. 129-130
30-Sep Spike Psarris, "Neptune Monument to Creation", *Creation Magazine,* December 2002-February 2003, pp. 22-24.
1-Oct Don DeYoung, "Space-Age Leaves", *Answers Magazine,* p.30-1, Jan-Mar 2010.
2-Oct Felice Gerwitz and Jill Whitlock, *Creation Anatomy,* 1996, p.15-16.
3-Oct George Mulfinger & Julia Mulfinger Orozco, *Christian Men of Science: Eleven Men Who Changed the World,* 2001, pp.71-97.
4-Oct Joanne E. De Jonge, *The Rustling Grass,* 1985, pp. 127-132.
5-Oct Joseph Mizzi and Michael Matthews, "The Amazing Cave People of Malta", *Creation Magazine,* December 2003, pp. 40-43. *Creation,* 26(1):40-43.
7-Oct Stuart Burgess, *The origin of man:The image of an ape or the image of God?,* p.24-28, 2004.
8-Oct Tom Vail, et al., *Your Guide to the Grand Canyon,* (Master Books, 2008), p. 157.
9-Oct 1. Lawrence Richards, *It Couldn't Just Happen,* pp. 27-29.
2. George Mulfinger and Julia Mulfinger Orozco, *Christian Men of Science,* 2001, pp. 118-145.
10-Oct Donald DeYoung, *Weather and the Bible,* 1992, p.60.
11-Oct Stuart Burgess, *The Origin of Man:The image of an ape or the image of God?,* p.124-133, 2004.
12-Oct Jeannie Fulbright, *Exploring Creation with Astronomy,* 2004, p. 55-57.
13-Oct Donald DeYoung and Derrik Hobbs, *Discovery Design: Searching out the Creator's Secrets,* (Master Books, 2009), pp. 45-47.
14-Oct 1. Michael Oard, "In the Footsteps of Giants", *Creation Magazine,* March 2003, p.10-12.
2. http://creation.com/images/pdfs/tj/j16_3/j16_3_5-7.pdf.
3. Tas Walker, "Dinosaur stumble preserved in Trackways, Utah,USA", *Journal of Creation,* August 2009, p. 4-5.
16-Oct 1. Richard and Tina Kleiss, *A Closer Look at the Evidence,* (Search for the Truth Pub., 2003), pp. 2,5.
2. George Mulfinger & Julia Mulfinger Orozco, *Christian Men of Science: Eleven Men Who Changed the World,* 2001, pp. 183-210.
17-Oct 1. Paula Weston, "Bats: sophistication in miniature", *Creation Magazine,* Dec. 1998, 21(1):28-31.
2. *Creation Moments,* "The Bat's Special Radar Design"
18-Oct Ken Ham, *Where Did Cain Get His Wife?* , 1997, pp.10-17.
19-Oct Dave Nutting, "Why the Mosquito?", *Alpha Omega Institute,* July-August 2006, newsletter 6.
20-Oct Jobe Martin, *The Evolution of a Creationist,* 1994, p. 193-195.
21-Oct 1. Paul S. Taylor, *The Great Dinosaur Mystery and the Bible,* p.37-43, 1989.
2. Ken Ham, *The Great Dinosaur Mystery Solved,* p.38, 162, 1998.
22-Oct 1. Andrew Snelling, "Creating Opals", *Creation Magazine,* December 1994, 14-17.
2. http://creation.com/creating-opals
23-Oct 1. Ron Samec, "The Heavens Declare a Young Solar System", *Answers Magazine,* Jan-March 2008, pp. 30-35.
2. http://www.answersingenesis.org/articles/am/v3/n1/heavens-declare-young-solar-system
24-Oct 1. Paula Weston, "Animal, vegetable and mineral: coral", *Creation Magazine,* December 2002-February 2003, pp.28-33.
2. http://creation.com/how-long-does-a-coral-reef-take-to-grow
3. Museum Guide: A Bible-based handbook to natural history museums, *AIG,* 2007, p.47.
25-Oct http://www.guardian.co.uk/science/2010/may/20/craig-venter-synthetic-life-form
26-Oct Steven Vogel, *Cats' Paws and Catapults: Mechanical Worlds of Nature and People* (262)
27-Oct http://www.answersingenesis.org/media/video/ondemand/seeing-eye/seeing-eye, Dr. David Menton *The Seeing Eye and Hearing Ear,* DVD.
28-Oct 1. Michael Oard, *The Missoula flood controversy and the Genesis Flood,* 2004, p.15-23.
2. Michael Oard, *Frozen in Time: The Woolly Mammoth, The Ice Age, and the Bible,* 2004, p.102-106.
29-Oct Paul Bartz, "The Double Life of the Hummingbird", *Letting God Create your Day,* p.178.
30-Oct 1. Dennis Dreves, "The Hummingbird: God's Tiny Miracle", *Creation Magazine,* December 1991, p.10-12.
2. John Ashton, *In Six Days: Why 50 Scientist Choose to Believe in Creation.*
3. http://www.answersingenesis.org/home/Area/isd/mcintosh.asp
31-Oct Donald B. DeYoung, *Weather and the Bible,* 1992, pp. 57-60. "

2-Nov John W. Oller, Jr., "More than PIE", *Answers Magazine*, p.52-55, April-June 2008.
4-Nov Bill Gothard, et al., *Character Sketches, Vol III*, Institute in Basic Youth Conflicts, 1985, p. 310.
5-Nov Andrew A. Snelling, "No Slow and Gradual Erosion", *Answers Magazine*, Jan-March 2009, 96-99.
6-Nov 1. Andrew Snelling, *Grand Canyon: Testimony to the Biblical Acc. of Earth's History*, 2009, DVD.
2. Andrew Snelling & Steven Austin, "Startling evidence for Noah's Flood Footprints and Sand "Dunes" in a Grand Canyon Sandstone!", *Creation Magazine*, December 1992 pp. 46-50.
7-Nov 1. www.physorg.com/news170010855.html
2. http://crev.info/ August 21, 2009
8-Nov Don DeYoung and John Whitcomb, *Our Created Moon*, 2003, p.80-81.
9-Nov Character Sketches, volume II, Institute in Basic Youth Conflict, Inc. 1978, p. 35.
10-Nov 1. Jason Lisle, *Taking Back Astronomy*, p.63-64, 2006.
2. Jason Lisle, *What does the Bible say about Astronomy*, p11-12, 2005.
11-Nov Don Batten, "Genesis contradictions?", *Creation Magazine*, September 1996, p.44-45.
12-Nov *Jonathan Park Goes to the Zoo*, DVD, 2008.
13-Nov 1. Ken Ham, *The New Answers Book*, 2006, pp.149-176.
2. Paul S. Taylor, *The Great Dinosaur Mystery and the Bible*, 1989, p.42.
14-Nov Philip Snow, *The Design and Origin of Birds*, p.18.
16-Nov 1. Stuart Burgess, *Hallmarks of Design*, 2004, p.137-140.
2. http://www.answersingenesis.org/articles/zoo/camel
18-Nov Donald DeYoung and Derrick Hobbs, *Discovery of Design*, p. 156-7, 2009.
20-Nov 1. Steve Cardno, "The Mystery of Ancient Man", *Creation Magazine*, March 1998, 20(2):10-14.
2. Donald Chittick, *The Puzzle of Ancient Man*, 2006, p. 157-159.
22-Nov Paula Weston, "Spectacular, Surprising Seals", *Creation ex nihilo*, September 2000, 22(4): 28-32.
23-Nov Thomas Heinze, "Did God Create Life? Ask a Protein", *Creation Magazine*, June 2006, pp. 50-52.
25-Nov http://www.answersingenesis.org/articles/am/v5/n2/elephant-video
26-Nov *Creation Magazine*, Sept. 1994.
27-Nov John Phillips, *Exploring Genesis: An Expository Commentary*, 1980, p35-37.
28-Nov Paul A. Bartz, *Creation Moments*, "The Bats Who Feed Trees".
29-Nov Jason Lisle, *Taking Back Astronomy*, 2006, pp.28-29.
30-Nov Dr. Jay L. Wile, *Exploring Creation with General Science*, 2000, pp.375-377.
1-Dec 1. Stuart Burgess, Hallmarks of Design, p.73-75, 95, 2004.
2. David Catchpoole, "Tale tale rates a 'fail' ", Creation Magazine, vol. 32, Issue 2, p.56, March 2009.
2-Dec Jeannie Fulbright, *Exploring Creation with Human Anatomy and Physiology*, pp.26-31, 2010.
3-Dec http://www.creationmoments.com/content/clean-blushing-fish
4-Dec http://garymoyers.blogspot.com/2006/12/swaddling-clothes.html
5-Dec Stuart Burgess, *The Origin of Man: The image of an ape or the image of God?*, 2004, p. 139-140.
6-Dec 1. Bill Johnson, "Thunderbirds did the American Indians see "winged dinosaurs"?, *Creation Magazine*, March 2002, pp.28-32.
2. Ken Ham, *The Great Dinosaur Mystery and the Bible*, 1998, pp.44-45.
7-Dec Donald DeYoung, *Astronomy and the Bible*, p. 54-55.
8-Dec Harun Yahya, *Atlas of Creation, vol. 1*, p.730-731, 2006.
9-Dec 1. Andrew A. Snelling, "Sand transported Cross Country", *Answers Magazine*, Vol.3 No.4 Oct-Dec 2008, p.96-99."
2. http://www.answersingenesis.org/articles/am/v3/n4/sand-transported
10-Dec Jerry Bergman, *The Dark Side of Charles Darwin*, (Master Books, 20111), pp. 194-200.
12-Dec David Catchpoole, "Bighorn Horns not so Big", *Creation* 32(4), 2010.
13-Dec 1. Dr. Jonathan Henry, *The Astronomy Book*, p.39., 1999.
2. Ken Ham, *The New Answers Book 2*, p. 50-51, 2008.
14-Dec John Hergenrather, "Noah's Long-Distance Travelers", *Creation*, June-August 2006, pp. 30-32.
15-Dec 1. http://www.bio.davidson.edu/Courses/Molbio/MolStudents/spring2005/Heiner/hemoglobin.html
2. James J. S. Johnson, *Survival of the Fitted: God's Providential Programming*, Acts and Facts, October 2010, pp. 17-18
16-Dec Ken Ham, Carl Wieland, Don Batten, *One Blood*, 1999, pp.68-72.
17-Dec Donald DeYoung and Derrick Hobbs, *Discovery of Design*, p. 158-9, 2009.
18-Dec John Sanford, Selection threshold severely constrains capture of beneficial mutations, *Biological Information - New Perspectives Conference*, Cornell University, May 2011.
19-Dec Brian Young, *Doubts about Creation? Not After This!*, 2008, pp. 241-245.
20-Dec Andrew Snelling, *Noah's Flood and the Earth's Age*, 2009.
21-Dec Dennis Peterson, *Unlocking the Mysteries of Creation*, p. 119, 2002.
22-Dec Jonathan Sarfati, *Creation Magazine*, "Super Shells", 27(3) June-August 2005, p.19.
23-Dec Privileged Planet, DVD
25-Dec Bill Gothard, et al., *Character Sketches, Vol III*, Institute in Basic Youth Conflicts, 1985, p. 147.
26-Dec Philip Snow, *The Design and Origin of Birds*, p.11, 108.
27-Dec John Bates, *A North Woods Companion: Fall and Winter*, p. 173.
29-Dec Tas Walker, "Mining Mountains in West Virginia", *Creation Magazine*, 12/08-2/09, pp. 48-49.
31-Dec Michael Oard, *Frozen in Time: The Woolly Mammoth, The Ice Age, and the Bible*, 2004, pp. 130-132, 141.

# INDEX BY TOPIC

# INDEX BY SUBJECT

## INDEX BY SUBJECT CONT.

INDEX BY SUBJECT CONT.

INDEX BY SUBJECT CONT.

# INDEX BY SCRIPTURE REFERENCE

# INDEX BY SCRIPTURE REFERENCE CONT.

## INDEX BY SCRIPTURE REFERENCE CONT.